The pH Diet

The pH Diet

Recharge Your Energy
Regain Your Figure
Restore Your Health

Bharti Vyas and Suzanne Le Quesne

Thorsons
An Imprint of HarperCollins*Publishers*
77–85 Fulham Palace Road,
Hammersmith, London W6 8JB

The website address is: www.thorsonselement.com

and *Thorsons* are trademarks of
HarperCollins*Publishers* Limited

Published by Thorsons 2004

1 3 5 7 9 10 8 6 4 2

A catalogue record of this book
is available from the British Library

ISBN 0 00 716511 0

Printed and bound in Great Britain by
Martins The Printers Limited, Berwick Upon Tweed

Contents

About the Authors

Bharti Vyas

For the past 25 years, Bharti Vyas has been a pioneering and influential voice in the field of health and beauty, recognizing that 'beauty on the outside, begins on the inside'. This has become the bedrock of her philosophy and where she has led, hundreds of beauty therapists now follow. Bharti's renowned BVM (Bharti Vyas Method) is taught all over the world. Bharti's knowledge of nutrition and its importance on our health and beauty comes from the knowledge of Ayurveda ('The Science of Life'), a healing system that has been used in India for thousands of years and that has been handed down through generations of her family. Its holistic approach treats the body, mind and spirit.

'My beauty philosophy is based on a simple principle: *beauty on the outside, begins on the inside.* To look your best you must feel your best, and to feel your best you must eat the right sort of foods in the right balance. Over the years I have seen thousands of clients and assessed not only their beauty requirements but also their general health from an Ayurvedic point of view. I believe that the level of acidity in the blood holds the key to many diseases in the body.

We are alkaline beings and, as such, our diets should be alkaline too. The foods used in the Ayurvedic system are mostly vegetables and fruits, which are mainly alkaline. In this book, I will share some of these principles with you.'

As part of her commitment to bringing her knowledge to the general public, Bharti has also written bestselling books *Beauty Wisdom*, *Simply Radiant*, *Simply Ayurveda* and *Fabulous Face*.

Travelling from all over the world to visit the Bharti Vyas Centre in London, her clients include royalty, politicians, celebrities and people from every profession.

Suzanne Le Quesne

Suzanne is a director of The Ludlow Clinic in Shropshire, UK, and has 14 years of clinical experience in nutrition.

'My health principles are those of optimum nutrition. Much disease occurs through people being exposed to an undesirable polluted environment. By providing an ideal biochemical environment within the body, we may be able to prevent and even cure such disease. To do this, we need to supply nutrients through foods and supplements, while eliminating 'anti-nutrients'.

'Food should be valued for its health-giving properties. The optimum-nutrition approach is fundamentally about empowering people to take charge of their own health.

Introduction

The pH Diet does exactly what it says on the cover. By following our programme, you'll recharge your energy, regain your figure and restore your health.

Does it sound too easy? If you're a diet aficionado, you may think so. There are countless weight loss and dietary programmes out there that you might have already tried – from lose-a-stone-in-two-weeks to high-protein or more conventional high carbohydrate/low fat diets. If you're reading this, maybe these diets haven't worked for you. They may have been too difficult to incorporate into your lifestyle, or made you feel deprived, so you lost your motivation to keep going. You need an eating plan that's easy and will work – and that's what *The pH Diet* delivers.

How does the pH diet work?

The pH Diet works by getting your body back to its naturally alkaline state. In three steps, or levels, you gradually cut down on acid-forming food and drinks, and eat more alkaline-forming foods.

Most of us have an inkling that too much coffee, red meat, sugar, dairy and processed foods, for example, are bad for us. In the pH Diet, these foods are discouraged because they're acid-forming. This means that when we've digested them, they leave an acidic residue in our bodies. All the food we eat 'burns' with oxygen in our cells to produce energy – our fuel. This digestion process generates an internal 'ash' that's acidic, alkaline or neutral. An alkaline or neutral ash is okay (our bodies are designed to be alkaline, not acidic). However, when acidic residue accumulates internally, it slows the body down – causing low energy, poor health and weight problems.

How does it benefit me?

By eating and drinking alkaline-forming foods, you release your body from the drudgery of coping with residual acid build-up. Rather than expend energy on fighting acidic toxins, your body can get back to its real job: fine-tuning your health and balancing your weight. Follow the pH diet and your energy levels will surge, you'll shed excess pounds, and will look and feel more vital than ever before. And there's no need to count calories, fat grams or points. You'll find straightforward diet plans (see pages 14, 23, 32) and food lists to follow (see page 39), so once you've grasped the acid/alkaline principle, you'll be well on your way to gaining and maintaining good health and a better figure.

The pH principle

The principle of the pH diet is that the most important aspect of a balanced, healthy body is our pH, or acid/alkaline balance.

pH is an abbreviation for 'potential of hydrogen'. It indicates the acidity or alkalinity of a solution, measured on a scale of 0–14. The lower the pH, the more acidic the solution; the higher the pH, the more alkaline the solution. When a solution is neither acid nor alkaline, it has a pH of 7, which is neutral.

You may have a vague idea about pH from school chemistry lessons, but pH is also used to market cosmetics and hair care products. 'pH-balanced' products are formulated to complement the pH of skin and hair. If you use products that are too acidic or too alkaline in nature, they may be damaging. If you wash your hair with detergent, for example, you wouldn't expect your locks to shine. Your hair would be clean but dry, dull and totally unmanageable. So choosing foods that are the wrong pH could leave you with the equivalent of a dry, dull and lifeless body. Visible signs will appear on the outside, reflected in poor-condition hair, nails and skin. On the inside, this imbalance shows up as symptoms that reflect the poor functioning of our internal organs (see the box on page xii to check any symptoms you may have).

Is the pH diet the same as food combining?

The answer is both yes and no. Like food combining, the pH diet is based on the concept of acid/alkaline imbalance as the cause of ill-health. This was first expounded in 1933 when William Howard Hay, a New York doctor, published his groundbreaking book, *A New Health Era*. He maintained that all disease is caused by autointoxication (self-poisoning) due to acid accumulation in the body. His work is still popular today and is now published as the Hay diet, also known as food combining.

However, the pH Diet is not a food-combining diet. When you follow the pH Diet, you can eat carbohydrates, proteins and fats together. This is because the pancreas produces three different digestive enzymes all at the same time, indicating that our bodies are designed to process different food groups simultaneously.

Is your diet draining your health?

Clinical trials have proved that an alkaline body is healthier than an acidic body. If you're regularly eating too many acid-forming foods, you will be more vulnerable to infection – from candida to frequent colds and flu. If you overindulge your love of meat, cheese, dairy, eggs, fish, alcohol and sugary and refined foods, for example, you are likely to have many minor, and some not so minor, symptoms. Symptoms are your body's way of getting your attention. By following the pH diet and eating more alkaline-forming foods, you can help reduce your symptoms and significantly improve your health.

Check your symptoms:
- Frequent infections, caused by a suppressed immune system: yeast infections, such as candida; parasitical infections; and bacterial and viral infections, such as colds and flu
- Low energy or chronic fatigue
- Aching muscles or joint pain; rheumatoid arthritis, osteoarthritis, gout or fibromyalgia
- Osteoporosis, weak, brittle bones; hip fractures or bone spurs
- Bladder or kidney problems, such as kidney stones
- Dull skin, brittle nails and hair
- Premature lines and wrinkles
- Liver or 'age' spots
- Acne, eczema or psoriasis
- Poor concentration or forgetfulness
- Excess weight, obesity
- Type II diabetes
- Mood swings

Stresses on the body

Our body tries its best to get rid of acidic residue left by acid-forming foods through urine, sweat and exhaled breath. However, our kidneys, skin and lungs can only cope with so much. They often become exhausted and cannot break down all the wastes from acid-forming foods, drinks and stimulants.

When this happens, what can't be processed has to be stored somewhere in the body. In order to live healthily, our blood and cells must always remain slightly alkaline. So the body, always pursuing survival, changes leftover acidic wastes into solid wastes and stores them.

Here are some examples of solidified acidic wastes:

- LDL cholesterol (the harmful cholesterol that can build up on artery walls)
- Adipose tissue (AKA fat)
- Uric acid (responsible for gout, kidney stones and gallstones).

The accumulation of these solid wastes can also be described as the ageing process and the cause of disease. When you eat abundant alkaline-forming foods you'll be able to excrete acid wastes far more effectively, and powerfully assist your whole body to function more efficiently. When your body is working in this way, weight loss is easy, symptoms disappear and good looks and health abound.

What is acidosis?

Most people who suffer from an unbalanced pH have too much acid in their bodies, a condition known as acidosis. This forces the body to 'borrow' minerals – including calcium, sodium, potassium and magnesium – from vital organs and bones to buffer the acid and safely remove it from the body. Because of this strain, the body can suffer severe and prolonged damage, and the condition may go undetected for years.

Acidosis is the foundation of many everyday symptoms like fatigue, poor skin, weak and brittle nails and difficulty in losing weight, as well as the many symptoms, illnesses and diseases listed in the box (see page xii). When you alkalize your body by following *The pH Diet*, you'll be able to restore and maintain your overall health and beauty.

So which foods are acid-forming?

One of my clients, with whom I'd been discussing acid- and alkaline-forming foods, expressed this important concern about what she should eat. 'What about lemon and lime?' she began. 'They're acidic yet they're antioxidants, and good for you.' This is a common assumption – that what tastes acidic stays acidic during digestion. However, the pH value of a food or drink isn't always the same as its acid- or alkaline-forming tendency in the body. It's what happens *after* we eat and drink that counts. 'Acidic' limes or lemons actually produce an alkaline residue in our bodies – the opposite of what we would expect. Likewise, meat doesn't taste acidic at all, but it leaves a very acidic residue in our bodies after digestion. So, like nearly all animal products, meat is very acid-forming.

All foods can be categorized as acid-forming, alkaline-forming or neutral. Water is neutral, against which all other foods and drinks are measured. To help you get your body back into pH balance, see the listing of acidic-forming foods and drinks to avoid (page 104) plus all the alkaline-forming super-foods you'll need to eat your way to great health, energy and weight loss.

But I already eat healthy foods ...

James, a client, had been feeling overweight and listless. In fact, his tiredness had been going on for so long that he admitted to me that he had almost become used to it. And the more tired he felt, the less he wanted to work out at the gym; and the more he used stimulants such as coffee and biscuits during the day to pep himself up. Otherwise, he thought he ate an excellent diet – lots of lean meat, potatoes, some vegetables.

On closer examination of James' eating habits, I explained to him how many of his favourite foods were acid-forming – and how his body could only fully assimilate nutrients when it's pH balanced. If your pH is too acidic you can eat healthy food packed with vitamins and minerals, yet get little or no health benefits. The goodness in food can't be absorbed in the gut because of the acidity there, so the nutrients are wasted. This helped to explain James' tiredness – he was nutrient-deficient because of his 'acid' diet. By changing to the pH Diet, he was able to break the tiredness cycle, cut out the stimulants and enjoy better health. He lost 5lbs in the process.

pH Values of Some Common Liquids

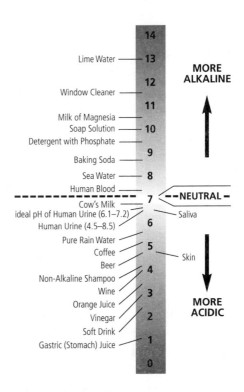

Getting the balance right

By eating enough alkaline-forming foods, you'll be able to establish and maintain a good acid/alkaline balance in your body. This is vital for good health, because acid-forming foods are inflammatory, whereas alkaline-forming foods are anti-inflammatory. So eating alkaline-forming, anti-inflammatory foods will give your body a tremendous boost. Acid-forming, inflammatory foods, however, may have only detrimental effects on the body. Below are examples of some common dietary habits that cause acid overload:

- A high-protein diet
- Eating lots of refined carbohydrates, such as bread and pasta
- Eating few vegetables and fruits
- Drinking high levels of alcohol

It's easy to see from the list above that none of these eating patterns provides a balanced diet. A balanced diet means getting the correct proportion of nutrients for health and vitality. The balance of carbohydrates, fats and proteins is important, but if you balance your diet between acid- and alkaline-forming foods by following the pH diet, you'll automatically get a balance of foods from the three major food groups. You won't feel deprived by forgoing all your carbs, or eliminating foods such as nuts because they're high in fat: with the pH diet, you'll regain your body's natural acid/alkaline balance, and get all the nutrition you need.

What about high-protein diets?

High-protein diets have become popular recently because they help people lose weight successfully. However, the long-term effects of diets such as these are not beneficial. Clinical studies have shown them to be a precursor to osteoporosis (brittle bone disease). In the short term, a high-protein diet will cause constipation, bad breath and low energy levels due to insufficient fibre and carbohydrate levels.

The urine of someone who regularly eats high-protein foods often gives an alkaline reading. However, this is misleading. When excess protein is consumed, the body releases calcium and magnesium from the bones and organs to neutralize the acids. As more and more minerals are taken from the bones, the weaker they become. A diet high in protein must therefore be considered with caution.

Starting the pH diet

The pH Diet is organized in three steps, or levels. All three levels are easy. In fact, they get easier as you move through them. They involve lifestyle as well as dietary changes.

In Level 1, you learn how to **reduce the toxic load** on your body. You do this by cutting back on the acid-forming liquids and anti-nutrients in your diet. These include:

- Milk
- Caffeine
- Nicotine
- Alcohol

Don't worry – you won't have to go cold turkey. You gradually reduce your intake of these toxins at a pace that suits you. You'll soon begin to feel the benefits, giving you the incentive to move up to Level 2.

In Level 2, you'll **increase your alkaline reserves**. This involves reducing wheat and including suggested alternatives, and introducing the 80 alkaline-forming super foods (see pages 43–87). There are lots of vegetables and vegetable juices in this part of the diet to rebuild the alkaline reserves. Vegetables are used for healing and are more alkaline-forming than the fruits used in Level 3. You can expect to be on Level 2 until alkaline reserves are in place. This could be six to eight weeks or longer, depending upon your previous levels, diet and lifestyle.

Level 3 concentrates on **maintaining the balance**. It explains how to eat a true alkaline-forming diet with less cooked food and more emphasis on raw foods. You'll also include more fruit and fruit juices in your diet. (Although fruits are mildly acidic, you can begin to eat them at this stage as you'll have built up more alkaline reserves, enabling your body to deal more efficiently with the acids they produce.)

We've included three **dietary plans** to help you choose your meals. There is one plan for each of the three levels, together with sheets for you to record your daily progress.

How long does the diet take?

You can spend as long as you like at each level of the programme. The longer you spend at each level, the more benefits you will get, and the more natural the programme will become to you. When you do go off track, which you probably will, simply start back at Level 1. One month at each level is beneficial, but knowing how keen you will be to start the programme, we suggest no less than a week on Level 1 before moving to Level 2. The longest possible time (four to six weeks minimum) should be spent on Level 2. Only when the alkaline reserves are in place can you move to Level 3.

Level 3 is the true pH Diet. It allows you to eat unlimited fruit and vegetables, and increase your intake of raw foods (up to 30 per cent of the diet). Because your overall health will have improved significantly and your body will be stronger by this level, you will be able to cope with more acid-forming foods. Although you can get away with the odd 'cheat' quite easily at Level 3, it is not recommended too often! You want to continue moving towards health and beauty and not backwards to ill-health and imbalance. Whenever you feel you need to, you can revert back to Level 2.

The more gently you treat your body by easing into the programme, and the longer you stay at each level, the less likely you are to suffer a 'healing crisis'. This is when the body is throwing off more toxins than you can eliminate, which may result in you feeling a little unwell.

You can check your pH status by testing your urine and saliva, but don't worry if you choose not to do this. It's not a compulsory part of this programme, and the diet will work equally well without it. However, should you start monitoring and recording your pH levels every day, you will feel more active on your journey to health. Testing your pH levels will give you regular updates on your progress – and it's fascinating and easy to do. *(For more information, see Chapter 4.)*

And finally

Food is one of life's great pleasures, so take your time and enjoy the programme. Your goal is to eat delicious foods that create alkaline reserves within your body. This will bring you better protection against disease, more energy, great-looking skin and a slimmer, healthier you. You can start right now.

CHAPTER ONE

Getting Started

The Principles of *The pH Diet*

Any new dietary programme should be entered into gently. The pH Diet is no exception. If your diet has been unbalanced or high in junk foods, or if you have been dieting on and off for a long period, don't make any sudden alterations to your eating habits. Such abrupt changes may result in withdrawal symptoms or a 'healing crisis', in which you may experience headaches or other unpleasant symptoms. We do not want this to happen to you, so we encourage you to take on the new programme slowly.

The pH Diet is not a race or a three-week wonder – it is an easy, flexible and healthy way of eating and enjoying food with the ultimate goal of regaining your health, losing weight and looking great. You should take the time to tune in to your body, its needs and responses, and take personal responsibility for it. Ultimately, no one knows more about you than you.

Flexibility and Benefits

There are three levels to the pH Diet. The amount of time you spend at each level is flexible, allowing you to build the programme around your lifestyle. However, the longer and more thoroughly you do each level, the greater the benefits will be. Doing it right is far more important than doing it quickly!

If you measure your pH status with the special paper as described in Chapter 4, it will be easier for you to decide when to move from Level 2 to Level 3. However, we recommend that you start at Level 1 and work through each level until symptoms improve and you feel and look better. Even if you have no current symptoms and your pH levels are ideal, prevention is better than cure. Following the programme can only benefit your long-term health.

Breakfast

Breakfast may be a challenge as it will probably be the meal with the most change. Level 1 introduces water and lemon juice on rising and reduces coffee or tea. Level 2 introduces millet flakes, Breakfast Rice and the '3&6 Mix'. Level 3 takes this further and introduces 'nut' milks while increasing fruit juices and fruit, maintaining the ever-decreasing intake of detrimental beverages.

The most important rule for breakfast is that you have some. So many adults and children leave home with no breakfast, snatching something unsuitable with no nutritional value on the way to work or school. A start to the day like this will only leave you feeling fatigued later in the day.

If you slip and have something you know to be extremely acid forming, such as a strong cup of coffee and a croissant or a bar of chocolate, drink more water to help dilute the toxins. Don't give yourself a hard time – just get back on track quickly. Start thinking of what your next meal will be. By thinking about alkaline super foods, you are more likely to eat something to help rebalance your system.

The pH Diet in a Nutshell
Level 1 – Reducing the toxic load
Level 2 – Increasing alkaline reserves
Level 3 – Maintaining the balance

Level 1 – Reducing the Toxic Load

Level 1 gets you started on banishing the acid wastes stored in your body. You need to start eliminating all the pollution that has built up in your body, especially the colon, from years of eating and drinking too many acid-forming foods and drinks. You do not, however, have to embark upon a sudden 'detox' regime. At this level you eat what you normally eat. The main emphasis here is liquid intake, not food intake. It makes good sense to reduce detrimental beverages and other anti-nutrients before embarking on changing your diet.

This is not a short-term programme, so you can do Level 1 for as long as you like, or as long as it takes for you to decrease your intake of harmful drinks and anti-nutrients. You need to clear out the old to make way for the new. Levels 1 and 2 will cleanse your body of impurities, normalize digestion and metabolism and regain alkaline balance. Leaving out the harmful beverages and anti-nutrients, before increasing beneficial foods and drinks, will give you an excellent start to reducing your weight, regaining your health and improving your looks.

Step 1: Cut Down on Harmful Drinks and Anti-nutrients

Many beverages contain caffeine. With every cup of coffee you drink, your body's calcium balance – the difference between calcium intake and excretion – becomes negative. If you drink more than a couple of cups (not mugs!) of coffee a day, the calcium loss can be substantial, especially if you are absorbing too little calcium in the first place. Drinking several cups of strong coffee or tea a day clogs up your body with toxins and slows you down mentally and physically.

Start now. Reduce your intake of the following drinks and anti-nutrients. Replace beverages with a healthier alternative from the 'Acid- and Alkaline-forming Drinks' table on page 8. Aim to reduce and eventually eliminate the anti-nutrients.

Beverages

- Coffee – including decaffeinated coffee
- Tea
- Alcohol – all wines, spirits and beers
- Fizzy drinks, including energy drinks and diet drinks
- Milkshakes – made from cow's milk
- Water flavoured with artificial sweeteners

Anti-nutrients

- Cigarettes
- Junk food – including sweets, ice cream, cakes, chocolates and crisps
- Processed foods – all ready-made meals
- 'Diet' foods or any 'low-fat', 'fat-free', or '99% fat-free' products (usually high in sugar, salt, preservatives and additives)
- Sugar

Step 2: Drink Water

Introduce water into your daily routine, starting each day with a glass of warm or cold water with added lemon or lime. The lemon or lime juice is very alkalizing and gives a good start to the day. Additionally, drink a glass of water on the hour every hour until 6pm. It may take a couple of weeks to get used to this, but it will become a habit. Still mineral water is more beneficial than carbonated water. Practise saying 'no' whenever you are offered an alcoholic drink, a coffee or anything you are trying to decrease or avoid – *you* are in control.

Step 3: Eat Breakfast

Have your usual breakfast. If you usually go without, choose something from the Level 2 breakfast list (*see page 22*). Breakfast is the most important meal of the day, so be sure to eat something every morning – no excuses. If you are keen to start reducing your toxic load, you can also begin to improve your breakfast by choosing something from Level 2 breakfasts.

Step 4: Enjoy Your Usual Lunch and Dinner

Have your usual lunch and evening meal. The objective at Level 1 is to reduce your toxic load by improving the quality of the liquids you are consuming and reducing the anti-nutrients. We concentrate on food more in Levels 2 and 3.

Beware Anti-nutrients

Just as nutrients maintain life, anti-nutrients are harmful to life. In following the pH Diet, you learn not only what to eat and drink for good health but also what not to eat and drink, which is equally important.

An anti-nutrient is any substance that stops beneficial nutrients being absorbed and used by the body, or any substance that promotes the excretion of a beneficial nutrient. Alcohol, cigarettes, man-made chemicals, pesticides, antibiotics and synthetic hormone residues can all be described as anti-nutrients. Even tap water may contain nitrates, lead or aluminium, all of which are anti-nutrients in their own right. Fried food, food cooked on a barbecue and burnt food can all be classed as anti-nutrients as they cause 'free-radical' damage to our cells. Free radicals are incomplete cells that attack healthy cells, causing damage that can lead to certain diseases, especially cancer and heart disease.

Many common medicines are also anti-nutrients. Aspirin, for example, irritates the gut wall, making it more permeable. The more anti-nutrients you take in, the more toxic and acidic you will be and the longer you should stay on Level 1, gradually reducing your intake of anti-nutrients over the coming weeks.

If you drink alcohol on most days – even just a glass of wine with dinner – introduce some alcohol-free days into your life. Start with one a week, and gradually build up. Alcohol is rich in calories and poor in nutrients. It also weakens your resolve and is likely to result in cheating, snacking or going off the rails completely. If you do not drink much alcohol but caffeine is your poison – in the form of coffee, tea or cola drinks – then try a caffeine-free day once in a while. As with alcohol, caffeinated drinks are deficient in nutrients, and there are many alternatives you may enjoy (see page 8). Again, start with one day a week and increase it gradually.

Nicotine is another anti-nutrient and should be reduced slowly. If you are having problems giving up smoking, you may find that alkalizing your body reduces cigarette cravings. A recent study has found that people smoke more when their diet is acidic.

Replacement Drinks

It will be easier to give up coffee, tea, hot chocolate, fizzy drinks and alcohol when you find an enjoyable alternative. There are a few ideas to help you in the table *(see page 8)*. Treat your body gently – and give up gradually. Giving up completely is the ultimate goal, but start by omitting the sugar and/or cow's milk, or make the drinks less strong, and drink less often. Your water intake will dilute the acid from the acid-forming drinks you are still having. So don't forget the water – *on the hour every hour until 6pm.*

Your aim is to reach the unlimited column of the table over the coming weeks and months. If you are starting from the most acidic column, drinking coffee, tea, milk and/or alcohol on most days, you can either reduce the number of times you consume the drink, drink it weaker, alternate it with water, or move on to the next column. Each time you move along a column you will be choosing a less acidic type of beverage. Over the coming weeks and months, work your way through the columns from left to right, making better choices and reducing the toxic load each time you move along a column. If your choices currently include tea and cow's milk and sugar, then gradually reduce the sugar and eventually the cow's milk before moving on to the next column.

Occasionally you may have a glass of wine, a cup of coffee or a milkshake. On these occasions, make sure you drink more water to detoxify the acidic drink and return to the alkaline-forming drinks and the unlimited column as soon as you can.

Gradually decrease your toxic load by moving along the columns until you reach the Unlimited column. This may take several months – persevere! As you move along the columns, your hair, skin, nails and eyes will look brighter, healthier and stronger.

Acid- and Alkaline-forming Drinks

DRINKS TO AVOID	SECOND CHOICE of acid-forming beverages – drink only occasionally	FIRST CHOICE of acid-forming beverages – drink sparingly	THIRD CHOICE of alkaline beverages – unlimited quantities	SECOND CHOICE of alkaline beverages – unlimited quantities	FIRST CHOICE of alkaline beverages – unlimited quantities	UNLIMITED
ACID FORMING DRINKS			ALKALINE FORMING DRINKS			
MOST ACID	ACID	LOWEST ACID	MOST ALKALINE	ALKALINE	LOWEST ALKALINE	NEUTRAL
Coffee Alcohol (wines, beers and spirits) Shop-bought 'carton' orange and fruit juices Cow's milk Milkshakes or smoothies made from cow's milk Cranberry juice Waters flavoured with artificial sweeteners Hot chocolate, instant or made with cow's milk	Decaffeinated coffee Tea with milk/sugar Weak coffee Spritzers (wine with soda water) Shop-bought 'carton' orange and fruit juices diluted by 50% with water	Black-leaf tea Earl Grey Most supermarket teas Soya Milk Nut Milks Buttermilk	Any fruit teas Home-made fruit juices Aqua Libra Soya Milk Milkshakes and smoothies made with soya milk	Green tea Herbal teas–fennel, chamomile, peppermint, nettle Dandelion coffee	Rooibos tea (red bush tea) Ginger tea with added lemon made with fresh ginger	Water Preferably bottled or filtered – you can add lemon and/or lime for taste Still water is the best choice. Carbonated water can reduce the absorption of some minerals.

Green Tea

The health benefits of green tea have been well established by many reputable studies. They contain powerful antioxidants, which have been shown to fight viruses, slow ageing, reduce high blood pressure, lower blood sugar, fight cancer and have an overall beneficial effect on health.

There are three types of tea: green, oolong and black. During the fermentation process, green tea is steamed, baked or pan-heated to prevent oxidation, thus the leaf remains green. Oolong tea is partially fermented. Oxidation is cut short so the leaves are black only on the edges. Black tea is fully fermented, producing black leaves. So why is green tea getting all the attention in the science world? It is mainly because of an antioxidant called EGCG that is preserved in green tea, but lost in oolong and black tea when fermented. Antioxidants are thought to prevent harmful free radicals. The highest amount of any known antioxidant is found in green tea. EGCG has been found to be 100 times more effective than vitamin C, and 25 times more effective than vitamin E, at neutralizing free radicals.

Ginger Tea

Ginger tea is warming and extremely good for settling the digestive system. It is best made from fresh ginger. Use about 2.5cm (1 inch) per person. Simply peel the ginger, slice it up and pour on boiling water. The longer you leave before drinking, the more beneficial it will be. Adding lemon will make it even more alkaline.

Herbal Teas

Correctly called tisanes, herbal teas are made from flowering plants without woody stems. Herbal infusions can include flowers, herbs, fruit and spices. Unlike all other types of tea, these infusions are caffeine-free. Examples are fennel, chamomile, nettle, peppermint, and rooibos (tea made from the red bush tree in South Africa, *see below*). These teas are alkaline forming, whereas fruit teas like apple, blackberry or raspberry may be mildly acid forming. However, we always have a choice and a mildly acid-forming fruit tea will be much more beneficial than a cup of extremely acid-forming coffee.

Tea

Most of the tea we drink is black tea. In fact, 94 per cent of all tea consumed is black. Although tea has approximately half the amount of caffeine as coffee, it should not be drunk excessively as it can discolour your teeth and cause insomnia. Drunk with meals, tea inhibits iron absorption by the body. Tea is without doubt a better choice than coffee, but ideally you should be looking to make an even better choice. Opt for green tea, ginger tea or rooibos tea, all high in antioxidants and beneficial to your health.

Rooibos Tea

The tea made from the red bush tree in South Africa is 100 per cent organic and has seemingly endless medicinal value. It really is nature's medicine. It is best drunk alone, but you can add lemon or ginger. Milk can be added, but takes away the refreshing taste of the tea. Wean yourself off cow's milk being added to any tea and choose soya milk as an alternative.

Rooibos tea:

- acts as an antioxidant that slows the ageing process
- prevents cancer and lowers the risk of cardiovascular disease
- is packed with flavonoids, antioxidants more powerful than vitamin C, green or black tea
- supports the digestive system and is anti-spasmodic, relieving stomach cramps and colic in babies

- helps manage allergies
- soothes skin irritations when applied directly to the affected area
- replenishes iron levels, so is useful for pregnant and menstruating women
- is calorie-free
- boosts the immune system
- aids health problems like insomnia, irritability, headaches, nervous tension and hypertension
- has a low tannin content (only 1–4 per cent)
- contains no artificial colours, additives or preservatives
- contains no caffeine, so can be drunk by pregnant women
- contains no oxalic acid, which prevents iron absorption
- contains copper, iron and potassium
- contains zinc for a healthy skin
- contains calcium, fluoride and manganese for strong bones and teeth
- contains magnesium for the nervous system.

Buttermilk

You might be surprised to learn that there is no butter in buttermilk and that it is quite a low-fat product. Old-fashioned, homemade buttermilk is the slightly sour, residual liquid that remains after butter is churned. The flavour of buttermilk is reminiscent of yoghurt, and most people prefer it well chilled. You'll find it slightly thicker in consistency than regular milk but not as heavy as cream. It takes 4.5 litres (1 gallon) of milk to yield 0.25 litres (half a pint) of true buttermilk. Nowadays, adding lactic acid bacteria culture to pasteurized skimmed milk makes most commercial buttermilk. As far as the pH Diet is concerned, buttermilk is an excellent product and can be substituted for milk in endless recipes. Buttermilk is only slightly acid forming whereas regular milk is very acid forming.

Cranberry Juice

Cranberry juice is one of very few fruit juices that have an acidifying effect on the body. For this reason, cranberry juice should be consumed only as a temporary measure, usually because there is a urinary tract infection. Its only purpose is to acidify the body quickly in order to provide symptomatic relief. Cranberry juice contains aromatic acids that are not metabolized by the body. These acids remain intact throughout the entire digestive process, acidifying the urine and bladder as they exit the body.

Cranberry juice may be helpful while the digestive system is adjusting to the new healthy-eating programme. It contains betaine hydrochloride with enzymes, which may be beneficial to very toxic people. If you have been following a high-protein diet, you may suffer from bladder infections. Cranberry juice will quickly remedy the situation. Cranberries have strong antibiotic properties with unusual abilities to prevent infectious bacteria sticking to cells lining the bladder and urinary tract. Cranberries also have anti-viral activity.

Why Water Is Vital

The water you drink is just as important as the food you eat. The majority of us will admit that we do not drink enough fluid, and that when we do drink, we make poor choices, such as alcohol, coffee, tea, hot chocolate and fizzy drinks. Water is of the utmost importance to becoming and remaining healthy, and for the hydration of the body, particularly the skin. Our bodies are 70 per cent water and our blood 94 per cent water. If we exist on polluted water, or too little water, imagine the devastation to our bodies.

The least desirable water for human consumption is the tap water provided by most cities and towns. Most city water contains chlorine, a powerful oxidizing agent that is not beneficial to the body and may in fact be harmful. Which water you drink is a matter of personal taste. Look at the ingredients list of the bottled water you buy. Many mineral waters have different levels of calcium and magnesium in them. Look for a balance of these minerals, ideally one that has a high ratio of magnesium to calcium as it is magnesium that keeps calcium in the bones. Avoid flavoured water as many have artificial sweeteners added. Artificial sweeteners have been proved to be carcinogenic (cancer forming) in animals. Choose still mineral water as carbonated mineral water may hinder beneficial minerals being absorbed. Become a water connoisseur!

Do Coffee and Tea Count Towards My Daily Water Intake?

No, unfortunately. Coffee and tea are diuretics and acid forming, and actually encourage your body to eliminate fluids. Even though they are liquid and contain water, they dehydrate the body rather than effectively providing the cells with the fluid they need.

The best kind of water for the body is the water contained in fruits and vegetables. When you change your diet to include predominantly fruits and vegetables, you will be surprised at how little additional water you need to keep your body satisfied and hydrated. However, it is still important to drink your daily allowance of water as well.

How Much Water Should I Drink?

It is recommended that you drink at least 2 litres (3^1/2 pints) of water daily, spread evenly over the day. This probably sounds like a lot of water, especially if you do not drink much at the moment, but as soon as you start drinking water at regular intervals, you will notice how much better you feel and look. Drinking water not only hydrates the inner body but also the outer body, giving the skin a lovely fresh glow.

If you are unused to drinking this amount of water, drink a glass on the hour, every hour until it becomes habit. If you do this from 8am until 6pm, you should have had your daily 2 litres (3¹/2 pints). If adding lemon or lime encourages you to drink this amount of water, then please do so – there is no limit to how much lemon you may add. Lemon will improve the taste and make the water alkalizing. You might own one of those watches that bleeps on the hour – now is the time to put it to good use as a reminder to drink your water.

Daily pH Diet Plan	Level 1
Reducing the toxic load	1–7 days (minimum)
On rising	A glass of cold or warm water with added lemon or lime.
On the hour and *every* hour until 6pm	A glass of still mineral water, with or without lemon/lime.
Breakfast	Your usual breakfast, using soya milk if you are having cereal, plus an improved beverage choice. If you usually have no breakfast, or want to make a start on reducing the toxic load from the first meal of the day, choose a breakfast from Level 2.
Mid-morning	Your usual mid-morning snack plus an improved beverage choice.
Lunch	Your usual lunch plus an improved beverage choice.
Mid-afternoon	Your usual mid-afternoon snack

	plus an improved beverage choice.
Dinner	No change in your choice of main meal but improved beverage choices.
Bedtime drink	Chamomile tea, rooibos tea, ginger tea or warm soya milk

Summary of Aims

- Water – on the hour every hour (285 millilitres/10 fluid ounces/$1^1/3$ cups per glass).
- Reduce intake of acid-forming beverages, including cow's milk.
- Start reducing anti-nutrients.

Level 1 – Reducing the Toxic Load

↓ Anti Nutrients

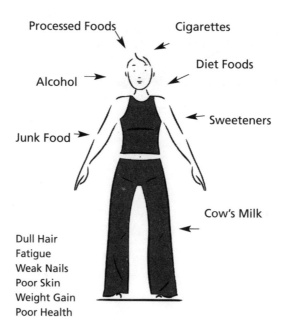

Processed Foods

Cigarettes

Diet Foods

Alcohol

Sweeteners

Junk Food

Cow's Milk

Dull Hair
Fatigue
Weak Nails
Poor Skin
Weight Gain
Poor Health

↑ Water

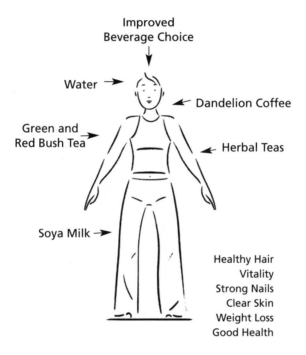

Improved
Beverage Choice

Water →

← Dandelion Coffee

Green and
Red Bush Tea →

← Herbal Teas

Soya Milk →

Healthy Hair
Vitality
Strong Nails
Clear Skin
Weight Loss
Good Health

Aims of Level 1 – Reduce Toxic Load
Reduce Anti-Nutrients and Improve Beverage Choices

REDUCE ANTI-NUTRIENTS	Give yourself a ✓ if you have decreased your intake of anti-nutrients. Make a note of the changes to monitor your progress.	✓	IMPROVE BEVERAGE CHOICE	Give yourself a ✓ if you have made improvement in your choice of beverage. Make a note of the changes to monitor your progress.	✓
CIGARETTES	From: To:		BREAKFAST	From: To:	
ALCOHOL	From: To:		MID-MORNING	From: To:	
SWEETENERS	From: To:		LUNCH	From: To:	
PROCESSED READY-MADE MEALS	From: To:		MID-AFTERNOON	From: To:	
DIET FOODS (low-fat, 99% fat-free)	From: To:		EVENING MEAL	From: To:	
JUNK FOOD (sweets, ice cream, cakes, crisps, chocolate)	From: To:		EVENING	From: To:	

WATER LOG Glass of water upon rising with lemon or lime?	7am	8am	9am	10am	11am	Noon	1pm	2pm	3pm	4pm	5pm	6pm	DAILY TOTAL ✓

BREAKFAST – Did you have breakfast today?

Did you use soya milk?

Scoring: 5–10 good   11–20 v.good   21+ excellent

Aims of Level 1 – Reduce Toxic Load
Reduce Anti-Nutrients and Improve Beverage Choices

REDUCE ANTI-NUTRIENTS	Give yourself a ✓ if you have decreased your intake of anti-nutrients. Make a note of the changes to monitor your progress.	✓
CIGARETTES	From: 20 a day To: 15 a day	✓
ALCOHOL	From: 1 glass red wine (pm) To: 1 glass Aqua Libra	✓
SWEETENERS	From: Don't use these To:	
PROCESSED READY-MADE MEALS	From: None today To:	
DIET FOODS (low-fat, 99% fat-free)	From: 99% fat-free biscuit To: A piece of fruit	✓
JUNK FOOD (sweets, ice cream, cakes, crisps, chocoalate)	From: None today To:	

IMPROVE BEVERAGE CHOICE	Give yourself a ✓ if you have made improvement in your choice of beverage. Make a note of the changes to monitor your progress.	✓
BREAKFAST	From: Coffee with milk and 2 sugars To: Coffee with milk and NO sugar	✓
MID-MORNING	From: Coffee with milk and 2 sugars To: Coffee with Milk and NO sugar	✓
LUNCH	From: Coffee with milk and 2 sugars To: Water	✓
MID-AFTERNOON	From: Tea with milk and 2 sugars To: Water	✓
EVENING MEAL	From: Tea with milk and 2 sugars To: Red bush tea (no milk or sugar)	✓
EVENING	From: Coffee with milk and 2 sugars To: Chamomile tea	✓

BREAKFAST – Did you have breakfast today? ✓
Did you use soya milk? ✓

WATER LOG — Glass of water upon rising with lemon or lime? ✓

7am	8am	9am	10am	11am	Noon	1pm	2pm	3pm	4pm	5pm	6pm	DAILY TOTAL
		✓	✓			✓	✓	✓	✓	✓		**18**

Example of a completed form

Level 2 – Increasing Alkaline Reserves

This level concentrates more on food, particularly foods that increase mineral reserves. These are green leafy vegetables and foods rich in essential fatty acids. During Level 2, you continue to reduce your toxic load by cutting down on wheat, substituting it with alternative types of grain. You are not eliminating wheat at this level, just reducing the amount you consume. In fact, you are not asked to give up any food, and you can continue with your regular diet. Level 2 stresses quality, not quantity, of food.

Step 1: Continue with Level 1

Each day, try to improve the drinks you consume and reduce anti-nutrients. Continue to drink water and lemon juice on rising for cleansing and alkalizing the system. Keep drinking 285 milli-litres/10 fluid ounces/1^1/3 cups water every hour, on the hour, until 6pm.

Step 2: Power Up Your Breakfast

Every morning, have 1 tablespoon of '3&6 Mix' (*see page 22*). Continue with your usual breakfast or choose one from the Level 2 breakfast list (*see page 22*).

Step 3: Introduce Vegetables and Vegetable Juices to Increase Alkaline Reserves

Shop-bought vegetable juice is acceptable on Level 2 if you really don't have time to make your own. Choose products with no added sugars or artificial ingredients. Some people find plain green juices difficult to stomach in the beginning. If so, consider adding carrot, apple or beetroot, which will sweeten the green vegetable juice. Gradually wean yourself off the carrot, apple or beetroot by upping the amount of greens. Visit a juice bar and experiment.

Step 4: Introduce Good Foods

Start introducing the 80 alkaline-forming foods and the 20 best acid-forming foods into your diet (*see Chapter 2*).

Step 5: Avoid Cow's Milk and Wheat

Avoid all cow's milk products, choosing soya alternatives. Restrict wheat products to once a day only.

Change and Improve
From ⟶ To

Wheat	Rye, oats, barley, millet, amaranth and corn
White processed bread	Brown bread
Brown bread	Wholegrain bread
Wholegrain bread	Burgen bread
Any cheese made from cow's milk	Goat's or sheep's cheese
Cow's milk	Soya milk or other alternatives such as rice milk

Level 2 Breakfasts

You may have your usual breakfast or choose from the following:

- Porridge made with water, served with 2 tbsp goat's/sheep's/soya yoghurt and '3&6 Mix' (*page 112*).
- Grapefruit (fresh or tinned), which is alkaline forming and a good source of fibre. It also eliminates acid wastes throughout your body, detoxifying your blood, tissues and digestive system. Top with fresh strawberries, raspberries or blackberries for added sweetness and alkalinity.
- Stewed figs or prunes served with a small tub of goat's/sheep's/soya yoghurt and '3&6 Mix' (*page 112*).
- Breakfast Rice with '3&6 Mix' (*page 112*).
- Millet flakes and '3&6 Mix' (*page 112*).
- Fresh fruit with '3&6 Mix' (*page 112*).
- Burgen bread is recommended. Burgen is a reduced-wheat bread made with soya flour and linseeds. Otherwise, buy the best bread you can find, preferably rye or a reduced-wheat product.
- 30g (1oz) nuts (almonds, hazelnuts, brazils or cashews) will add protein for balance.

The '3&6 Mix'

The '3&6 Mix' is a mix of seeds giving a balance of the omega-3 and omega-6 essential fatty acids. These fatty acids are essential to life and must be taken in the diet on a daily basis.

The omega-6 essential fats make hormone-like substances called prostaglandins. These are involved in inflammatory reactions and in the stickiness of the blood. If you lack omega-6 fats, you may suffer from eczema, allergies, premenstrual tension and even hyperactivity. The omega-3 essential fats make a different prostaglandin, which is involved in keeping the heart and arteries healthy, and again influencing the formation of blood clots. We need both types every day for optimal health. Although these essential fats can be obtained from other foods, the '3&6 Mix' is a quick and easy way to ensure you are getting your quota every day.

In the pH Diet we are going to use half the seeds as they are, sprinkled over vegetables or added to sandwiches, and the other half ground in a simple coffee grinder, added to soups or cereals.

Organic flaxseeds make up 55 per cent of the mix. In addition to being a beneficial source of omega-3 fatty acids, flaxseeds have water-soluble fibres, which are very effective at relieving constipation. Many people are deficient in omega-3 fatty acids, and getting your daily requirement through the seed mix is easy. Organic sesame seeds comprise 15 per cent of the mix, as do organic pumpkin seeds and sunflower seeds.

The '3&6 Mix' is an easy way to obtain these 'essential for life' oils with very little effort. You can follow the recipe described on page 112.

Mid-morning Juice

We recommended that you have a mid-morning vegetable juice. If this is not practical, then have an improved beverage and take the vegetable juice at some other time during the day. The important thing is that you have it at some time. These juices and extra portions of green leafy vegetables will increase the alkaline reserves in your body.

Choose from the following delicious selection, recipes for which can be found in Chapter 3:

- Mixed Vegetable Juice
- Watermelon Juice
- Alkaline Special

- The Waldorf
- Potassium Special
- Super Skin Vegetable Juice
- Carrot and Apple Juice
- Acne Cleanser
- Carrot and Parsley Juice
- Summer Cocktail
- Nail Strengthener
- Super Skin Fruit Juice
- Bone Builder
- Liver Cleanser
- Melon Juice
- Wrinkle Smoother
- Body Cleanser
- Green Power Juice

Daily pH Diet Plan	**Level 2**
Increasing Alkaline Reserves	1–2 months
On rising	A glass of cold or warm water with added lemon or lime.
On the hour and **every** hour until 6pm	A glass of still mineral water, with or without lemon or lime.
Breakfast	Your usual breakfast, using an alternative to cow's milk if you are having cereals, or a Level 2 breakfast, plus an improved beverage choice.

Mid-morning

V8 Juice (a low-calorie, tinned vegetable juice with no added sugar or preservatives), tomato juice or home-made vegetable juice.

Lunch

Lunch should include additional vegetables, such as a salad or a lightly cooked green leafy vegetable, plus a small portion of low-fat, good-quality protein and essential fatty acids (nuts and seeds or '3&6 Mix'). A Greek salad or Caesar salad would fit the bill nicely.

Mid-afternoon

Fruit or fruit juice (no added sweeteners if shop bought).

Dinner

Vegetable soup (preferably home made) or an additional serving of green vegetables, plus a small portion of protein and essential fatty acids.

Bedtime drink

Chamomile tea, rooibos tea, ginger tea or warm soya milk.

Your Goals

- Continue with Level 1.
- '3&6 Mix' daily.
- Wheat – once a day *maximum*.
- Four servings of green leafy vegetables every day.
- One serving of fruit or fruit juice daily.

Level 2 – Increasing Alkaline Reserves

↓ Anti Nutrients

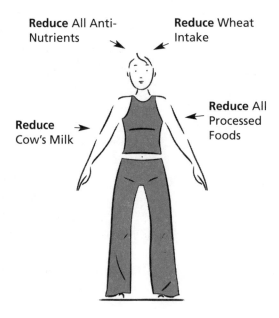

Reduce All Anti-Nutrients

Reduce Wheat Intake

Reduce Cow's Milk

Reduce All Processed Foods

↑ Alkaline Foods

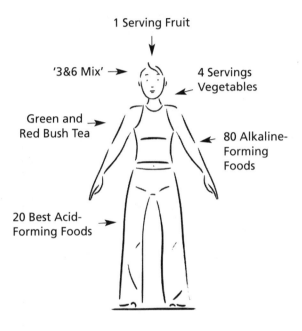

1 Serving Fruit

'3&6 Mix' →

4 Servings Vegetables

Green and Red Bush Tea →

80 Alkaline-Forming Foods

20 Best Acid-Forming Foods →

Aims of Level 2 – Increasing Alkaline Reserves
Food – Quality Not Quantity – Increasing Vegetables

	Give yourself a ✓ if you have decreased your intake of anti-nutrients. Make a note of the changes to monitor your progress.	✓ CONTINUE TO IMPROVE BEVERAGE CHOICE	MAIN MEAL IMPROVEMENTS	✓ Give yourself a ✓ if you have made an improvement in your choice of meal. Changes include reducing wheat, substituting dairy and increasing vegetables
CONTINUE TO REDUCE ANTI-NUTRIENTS				
CIGARETTES	From: To:	BREAKFAST	BREAKFAST	'3&6 Mix'? Non-wheat breakfast cereal? Burgen bread (reduced wheat) or non-wheat bread? Soya milk?
ALCOHOL	From: To:	MID-MORNING	MID-MORNING	Vegetable Juice?
SWEETENERS	From: To:	LUNCH	LUNCH	Extra serving of vegetables or salad? Small portion of low-fat protein? Nuts and seeds or '3&6 Mix'?
PROCESSED READY-MADE MEALS	From: To:	MID-AFTERNOON	MID-AFTERNOON	1 serving of fruit? (if pref, this can be eaten at breakfast or after an evening meal – at Level 2 only 1 serving per day is allowed)
DIET FOODS (low-fat, 99% fat-free)	From: To:	EVENING MEAL	EVENING MEAL	Vegetable soup – home made or shop bought? Vegetables to make up to 4 servings per day? Portion of good-quality, low-fat protein?
JUNK FOOD (sweets, ice cream, cakes, crisps, chocolate)	From: To:	EVENING	EVENING	Your evening meal should have satisfied your hunger. If you still require food choose a vegetable snack such as crudités or your fruit allowance.

WATER LOG Glass of water upon rising with lemon or lime?	7am	8am	9am	10am	11am	Noon	1pm	2pm	3pm	4pm	5pm	6pm	DAILY TOTAL ✓

BREAKFAST – Did you have breakfast today?
Did you use soya milk?
Wheat – max. 1 serving per day
Fruit – max. 1 serving per day
Vegetables – max. 4 servings per day

Scoring: 5–10 good 11–20 v.good 21+ excellent

Aims of Level 2 – Increasing Alkaline Reserves
Food – Quality Not Quantity – Increasing Vegetables

CONTINUE TO REDUCE ANTI-NUTRIENTS ✓	CONTINUE TO IMPROVE BEVERAGE CHOICE ✓	MAIN MEAL IMPROVEMENTS ✓
Give yourself a ✓ if you have decreased your intake of anti-nutrients. Make a note of the changes to monitor your progress.		Give yourself a ✓ if you have made an improvement in your choice of meal. Changes include reducing wheat, substituting dairy and increasing vegetables
CIGARETTES — From: 15 a day / To: 5 a day	✓ BREAKFAST	✓ BREAKFAST — '3&6 Mix'? Non-wheat breakfast cereal? Burgen bread (reduced wheat) or non-wheat bread? Soya milk? ✓ ✓
ALCOHOL — From: 3 glasses of red wine / To: 2 white wine spritzers	✓ MID-MORNING	✓ MID-MORNING — Vegetable Juice? ✓
SWEETENERS — From: Don't take these / To:	✓ LUNCH	✓ LUNCH — Extra serving of vegetables or salad? Small portion of low-fat protein? Nuts and seeds or '3&6' Mix? ✓ ✓
PROCESSED READY-MADE MEALS — From: Pizza at lunchtime / To: Greek salad	✓ MID-AFTERNOON	✓ MID-AFTERNOON — 1 serving of fruit? (if pref., this can be eaten at breakfast or after an evening meal – at Level 2 only 1 serving per day is allowed) ✓
DIET FOODS (low-fat, 99% fat-free) — From: None today / To:	✓ EVENING MEAL	✓ EVENING MEAL — Vegetable soup – home made or shop bought? Vegetables to make up to 4 servings per day? Portion of good-quality, low-fat protein? ✓ ✓
JUNK FOOD (sweets, ice cream, cakes, crisps, chocolate) — From: A chocolate bar mid-morning / To: 6 almonds	✓ EVENING	✓ EVENING — Your evening meal should have satisfied your hunger. If you still require food choose a vegetable snack such as crudités or your fruit allowance.

WATER LOG	7am	8am	9am	10am	11am	Noon	1pm	2pm	3pm	4pm	5pm	6pm	DAILY TOTAL
Glass of water upon rising with lemon or lime?			✓		✓	✓	✓	✓	✓	✓	✓		30

BREAKFAST – Did you have breakfast today? ✓
Did you use soya milk? ✓
Wheat – max. 1 serving per day ✓
Fruit – max. 1 serving per day ✓
Vegetables – max. 4 servings per day ✓

Example of a completed form

Level 3: Maintaining the Balance

Now the body is replenished with adequate alkaline reserves from the vegetables and vegetable juices, you can increase the amount of fruits and fruit juices you consume. The body is now better equipped to deal with their slightly more acid residue. To be sure you are ready to move to Level 3, it is useful to test the pH of your urine and saliva (*see Chapter 4*). 75 per cent of foods in Level 3 consist of alkaline-forming fruit and vegetables, and 25 per cent of acid-forming foods like grains, nuts, seeds, meat, fish, organic eggs and poultry. On this level, 30 per cent of the alkaline-forming foods should be raw.

Step 1: Continue with Levels 1 and 2

Keep improving beverage choices, reducing anti-nutrients and trying out alternatives.

Step 2: More Fruit

Increase the amount of fruit and fruit juices in your diet.

Step 3: More Raw Foods

You should increase the amount of salads and vegetable juices you consume.

Step 4: Get the Balance Right

The 80 alkaline-forming foods should make up 80 per cent of your diet, of which 30 per cent should be raw. The rest of your diet should consist of the 20 best acid-forming foods (*see Chapter 2*).

Step 5: Cut the Junk

There should now be no processed foods, anti-nutrients, additives or preservatives in your diet.

The Importance of Protein

Please remember that the pH Diet does not restrict protein. However, only 15 per cent of your daily calorie intake should come from this food source. Protein is necessary for the maintenance of cells, function of metabolism, manufacture of hormones and neurotransmitters and ability to bind toxins. Excess protein, however, can make the body too acidic. This can interfere with cellular metabolism resulting in a loss of vital minerals, reduced energy, poor digestion and reduced ability to clear toxins.

Did You Know ...
That it takes 20 times as much alkaline to neutralize 1 part acid. For example, you would need to drink 20 cups of water (which is alkaline) to neutralize 1 cup of acid-forming milk, alcohol, coffee or cola.

Alkaline-forming Super Foods

There are 80 alkaline-forming super foods listed in Chapter 2 and they are all great for your health. Here are our seven favourites:

- Avocados
- Canaloupe melons
- Grapefruit
- Green beans
- Limes and lemons
- Tomatoes
- Water – although neutral, it is classed as a super drink when it has added lemon or lime juice

Fruit

All high-sugar fruits are slightly acid forming, which is why they are increased in Level 3, after alkaline reserves have been built up. This list will show you just how much sugar fruit contains:

High-sugar Fruits – Mildly Acidic	Low-sugar Fruits – Alkalizing
Pineapples: 28% sugar	Avocados: 2% sugar
Ripe bananas: 25% sugar	Tomatoes: 3% sugar
Honeydew melons: 21% sugar	Lemons: 3% sugar
Apples: 15% sugar	Cantaloupe melons: 5% sugar
Oranges: 12% sugar	Non-sweet grapefruit: 5% sugar
Strawberries: 11% sugar	
Watermelons: 9% sugar	

Alkaline or Acid?

Alkaline-forming foods are usually considered beneficial while acid-forming foods are seen as detrimental. However, there are always grey areas and exceptions to the rule. There is a difference of opinion as to which foods are acid forming and which are alkaline forming.

Bananas, avocados, asparagus, artichokes and spinach are considered by Zen Buddhists to be acid producing, while Western scientists believe them to be alkaline because, when burned, they leave mostly alkaline mineral ashes.

Because many acid-forming fruits and vegetables are effective cleansers of the body's acid wastes, they are classed as beneficial. The juice of carrots and beets, with their high percentage of acid-forming sulphur and phosphorus, effectively cleans out the acid wastes from the liver, kidneys and bladder. The juice of cabbage, which is high in acid chlorine and sulphur, cleanses the acid wastes adhering to the mucous membranes of the stomach and intestinal tract. An excellent remedy for gum disease and infections in general is the highly acidic vitamin C.

Alkaline minerals are also effective cleansers. Potassium, calcium, sodium and magnesium reduce excess acidity in all the organs of the body. Indeed, acid and alkaline minerals act together to cleanse the body, just as a combined solution of vinegar (acid) and bicarbonate of soda (alkaline) makes an excellent household cleaner.

We need to make a distinction between acids that are toxic and cause the body's organs to degenerate and those that rid the body of acid wastes and, by doing so, prevent degenerative disease. We cannot just assume that foods high in acid-forming minerals will add to the store of acid wastes in the body and therefore must be avoided. The more alkaline reserves we have, the better our bodies will be able to handle the acid-forming foods. Ripe bananas are slightly more acid forming than unripe bananas, but ripe bananas are recommended because they are much easier to digest – but eat only in moderation as they also have a high glycaemic index, which can raise blood sugar levels rapidly.

Did You Know ...

That lemons, limes, tomatoes and avocados are low in sugar and high in water content. They produce very little acid residue and are highly alkalizing.

Soaking Nuts and Seeds

Some foods become far more effective when soaked. Nuts and seeds are such foods. Soaking them activates their enzymes and partially digests the protein, making all the nutrients they contain readily available to the body. Soaking also makes small seeds, such as sesame and flax, easier to chew and therefore to digest.

Place nuts and seeds in a container, cover with water to 2.5–5cm (1–2 inches) above the top of the nuts, and place in the refrigerator for an hour or two or, for almonds, overnight. They will plump up, absorbing the water and the oxygen in the water. Then they will be ready to eat and enjoy. Rinse them and change the water ever day. Keep them totally submerged. Eat within two days to prevent mould forming under the skin of the nuts.

Daily pH Diet Plan Maintaining the Balance	Level 3
On rising	A glass of cold or warm water with added lemon or lime.
On the hour and *every* hour until 6pm	A glass of still mineral water, with or without lemon or lime.
Breakfast	Fruit Breakfast and '3&6 Mix'.
Mid-morning	Fruit, fruit juice, vegetable juice or water.

Lunch	A large mixed salad or a lightly cooked green leafy vegetable plus a small portion of low-fat, good-quality protein and essential fatty acids ('3&6 Mix'). A salade niçoise, or mozzarella and tomato salad drizzled with olive oil, or half an avocado with prawns and a large salad, would be perfect.
Mid-afternoon	Fruit, fruit juice, vegetable juice or water.
Dinner	Vegetable soup (preferably home made) or an additional serving of green vegetables, plus a small portion of protein and essential fatty acids. You could have a small jacket potato (eat the skin) for lunch or dinner.
Bedtime drink	Chamomile tea, rooibos tea, ginger tea or warm soya milk.

Your Goals

- 80 per cent alkaline-forming foods (of which 30 per cent should be raw).
- 20 per cent acid-forming foods.
- Wheat – three times a week *maximum*.
- Six servings of fruit and vegetables *daily*.
- Experiment – introduce a 'new' food each week.
- Maintain the goals of Levels 1 and 2.

Level 3 – Maintaining the Balance

↓ Anti Nutrients

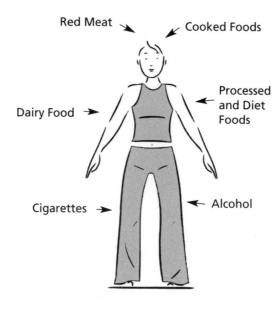

↑ Alkaline-Forming Foods

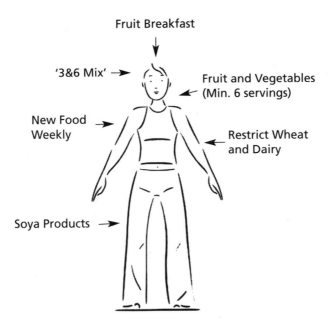

Fruit Breakfast

'3&6 Mix' →

Fruit and Vegetables
(Min. 6 servings)

New Food
Weekly

Restrict Wheat
and Dairy

Soya Products →

Aims of Level 3 – Maintaining the Balance
Increasing Fruits and Vegetables – Reducing Wheat and Dairy

	CONTINUE TO REDUCE ANTI-NUTRIENTS — Give yourself a ✓ if you have continued to decrease your intake of anti-nutrients from last week.	CONTINUE IMPROVEMENTS TO BEVERAGE CHOICE ✓	MAIN MEAL IMPROVEMENTS — Give yourself a ✓ if you have made an improvement in your choice of meal. Changes include reducing wheat, substituting dairy and increasing fruits & raw vegetables. ✓
CIGARETTES	From: / To:	BREAKFAST	BREAKFAST — '3&6 Mix'? Fruit Breakfast? Soya yoghurt?
ALCOHOL	From: / To:	MID-MORNING	MID-MORNING — Vegetable or Fruit Juice?
SWEETENERS	From: / To:	LUNCH	LUNCH — Large mixed salad or lightly cooked veg(s)? Small portion of low-fat protein? Nuts and seeds or '3&6 Mix'? Foods from list on page 44?
PROCESSED READY-MADE MEALS	From: / To:	MID-AFTERNOON	MID-AFTERNOON — Fruit, fruit juice, vegetable juice or water?
DIET FOODS (low-fat, 99% fat-free)	From: / To:	EVENING MEAL	EVENING MEAL — Vegetable soup – home made or shop bought? Foods from list on page 44 or a portion of good-quality, low-fat protein? Vegetables and fruit to reach daily minimum?
JUNK FOOD (sweets, ice cream, cakes, crisps, chocolate)	From: / To:	EVENING	EVENING — Your evening meal should have satisfied your hunger. If you still require food choose a vegetable snack such as crudités or your fruit allowance.

FRUIT BREAKFAST ?
Wheat today?
Max. 3 servings per week Fruit & Vegetables
Min. 6 servings every day
NEW FOOD TODAY?

WATER LOG — Glass of water upon rising with lemon or lime?	7am	8am	9am	10am	11am	Noon	1pm	2pm	3pm	4pm	5pm	6pm	DAILY TOTAL ✓

Scoring: 5–10 good 11–20 v.good 21+ excellent

Aims of Level 3 – Maintaining the Balance
Increasing Fruits and Vegetables – Reducing Wheat and Dairy

	Give yourself a ✓ if you have continued to decrease your intake of anti-nutrients from last week.	CONTINUE IMPROVEMENTS TO BEVERAGE CHOICE ✓	MAIN MEAL IMPROVEMENTS ✓	Give yourself a ✓ if you have made an improvement in your choice of meal. Changes include reducing wheat, substituting dairy and increasing fruits & raw vegetables.
CONTINUE TO REDUCE ANTI-NUTRIENTS ✓				
CIGARETTES	From: 5 a day To: NON-SMOKER !!	BREAKFAST ✓	BREAKFAST ✓	'3&6 Mix'? ✓ / Fruit Breakfast? ✓ / Soya yoghurt?
ALCOHOL	From: 2 white wine spritzers To: None	MID-MORNING ✓	MID-MORNING ✓	Vegetable or Fruit Juice? ✓
SWEETENERS	From: Don't take these To:	LUNCH	LUNCH ✓	Large mixed salad or lightly cooked veg(s)? ✓ / Small portion of low-fat protein? ✓ / Nuts and seeds or '3&6 Mix'? ✓ / Foods from list on page 44?
PROCESSED READY-MADE MEALS	From: None To:	MID-AFTERNOON	MID-AFTERNOON ✓	Fruit, fruit juice, vegetable juice or water? ✓
DIET FOODS (low-fat, 99% fat-free)	From: None To:	EVENING MEAL	EVENING MEAL ✓	Vegetable soup – home made or shop bought? ✓ / Foods from list on page 44 or a portion of good-quality, low-fat protein? ✓ / Vegetables and fruit to reach daily minimum? ✓
JUNK FOOD (sweets, ice cream, cakes, crisps, chocolate)	From: None To:	EVENING	EVENING ✓	Your evening meal should have satisfied your hunger. If you still require food choose a vegetable snack such as crudités or your fruit allowance.

WATER LOG Glass of water upon rising with lemon or lime? ✓	7am	8am	9am	10am	11am	Noon	1pm	2pm	3pm	4pm	5pm	6pm	DAILY TOTAL ✓
			✓		✓	✓	✓	✓	✓	✓	✓	✓	33

FRUIT BREAKFAST ? ✓
Wheat free today?
6 servings Fruit & Vegetables ✓
NEW FOOD TODAY? ✓ Artichoke

Example of a completed form

A Word for Vegetarians

Peas, beans, legumes, rice and other complex carbohydrates do contain proteins but are normally deficient in one or another (lysine or methionine). To ensure you get all the essential amino acids, a general rule of thumb is to have a pulse with a grain. For example, beans on toast would give you the pulse from the beans and the grain from the bread. The best course of action is to have as much variety in your diet as possible; include the '3&6 Mix' for the essential fatty acids; and soya, nuts and 'nut milks' as main sources of protein. With an abundance of green leafy vegetables, soya milk, goat's or sheep's milk and cheese, you will also receive sufficient calcium.

What About Supplements?

If your diet is balanced across all the major food groups, and you are eating enough good-quality fruits and vegetables, then there should, in theory, be no need for nutritional supplements. However, many people believe that due to the fast way our food is grown and harvested, it contains fewer nutrients than it did just 50 years ago. Supplements are a huge area and beyond the scope of this book. However, here are some recommendations for those who do choose to supplement their diet:

- Always take a multivitamin as opposed to individual vitamins or minerals. If you take just one vitamin or mineral, there is a good chance you will make a deficiency in another.
- Always take the B-vitamins as a B-complex, and take them in the morning. Vitamins give vitality whereas minerals are natural calming agents and should be taken in the evening.
- A good antioxidant is often needed, especially if there is a history of cancer in the family. Again, take as a 'multi-antioxidant', one that contains vitamins A, C, E and the

mineral selenium, especially formulated to work together. Pine bark extract is one of the most valuable antioxidants. It helps to bind up acidity, thereby reducing inflammation (aches and pains) in the body. It has been shown to bind directly with the body's connective tissue, maintaining and repairing it.

- Always take calcium with magnesium, never calcium on its own. Take it in the evenings to make you feel relaxed and calm.
- Digestive enzymes are invaluable for people with digestive disorders. Again, a 'multi-digestive' enzyme will contain all the enzymes needed to break down the different foods groups.

CHAPTER TWO

The Best Foods for the Diet

In this chapter we will look in detail at 80 alkaline-forming super foods and the 20 best acid-forming foods. Once you start incorporating these wonderful foods into your life, and seeing and feeling the benefits of eating well, you will wonder how you ever managed without them. The foods are listed alphabetically for ease.

Free Radicals and Antioxidants

These two terms crop up rather a lot in this chapter.

Free radicals are the bad guys, incomplete cells that attack healthy cells, causing damage that can lead to certain diseases, especially cancer and heart disease. Free radicals are major players in the build-up of cholesterol in the arteries that leads to atherosclerosis (hardened arteries) and heart disease; the nerve and blood vessel damage seen in diabetes; the cloudy lenses of cataracts; the joint pain and damage seen in osteoarthritis and rheumatoid arthritis; and the wheezing and airway tightening of asthma.

Antioxidants are the good guys. These powerful substances travel through the body neutralizing acidity and free radicals.

	ALKALINE			ACID
ALFALFA	CARROTS	KELP	POTATO SKINS	BREAD — Burgen bread, Bharti's Bread, corn and rye breads
ALMONDS	CAULIFLOWER	KIWI FRUIT	PUMPKIN	CHERRIES
AMARANTH	CELERY	LEEKS	QUINOA	CHICKEN
APPLES	CHESTNUTS	LEMONS & LIMES	RADISHES	COTTAGE CHEESE
APRICOTS	CHICORY	LETTUCE	RAISINS	EGGS — organic
ARTICHOKES	CHIVES	MANGOS	SORREL	FISH — salmon, cod, sardines, haddock, tuna or any cold-water fatty fish
ASPARAGUS (ripe)	COURGETTES/ZUCCHINI	MILLET	SOYA BEANS — Soya cheese & Soya milk	KIDNEY BEANS
AUBERGINES (EGGPLANTS)	CUCUMBER	MUSHROOMS	SPINACH	LAMB
AVOCADOS	ENDIVE	OKRA	STRAWBERRIES	LENTILS
BANANAS (ripe)	FENNEL	OLIVE OIL & OLIVES	SWEETCORN (fresh only)	LIMA BEANS (dried)
BEAN SPROUTS	FIGS	ONIONS	SWEET POTATO	NUTS
BEETROOT	FLAXSEED OIL	ORANGES	TAMARIND	OATS
BLACKBERRIES	GARLIC	PAPAYA	TANGERINES	PLUMS
BROCCOLI	GINGER	PARSLEY	TOFU	PUMPKIN SEEDS
BRUSSELS SPROUTS	GOAT'S CHEESE & MILK	PARSNIPS	TOMATOES	RICE — brown
BUCKWHEAT	GRAPEFRUIT	PEACHES	WATERCRESS	RYE
CABBAGE, SAUERKRAUT & BOK CHOY	GRAPES	PEARS	WATERMELON	SUNFLOWER SEEDS
CANOLA OIL	GREEN BEANS	PEAS	WHEATGRASS	TORTILLAS
CANTALOUPE MELON	HONEY (raw)	PEPPERS	WILD RICE	TURKEY
CAROB	KALE	PINEAPPLES	WINTER SQUASH	YOGHURT

The 80 Top Alkaline-forming Foods and the 20 Best Acid-forming Foods

The 80 Top Alkaline-forming Foods

These foods should make up 80 per cent of your daily diet. Not only are all the foods alkaline forming, they have many health-giving properties in their own right. You can eat as much or as many of them as you like.

Alfalfa

Usually eaten as sprouted seeds, alfalfa can be bought from super-markets and health-food shops. Alfalfa contains the fat-soluble vitamins A, D, E and K and some of the B-vitamins. It is high in protein, and contains a good ratio of calcium to magnesium, plus phosphorus, iron, potassium, chlorine, sodium, silicon and other trace elements. It has eight enzymes known to promote the chemical reactions that enable food to be assimilated properly within the body.

Alfalfa aids stomach ailments, ulcerous conditions, and the pain and stiffness of arthritis. It may eliminate retained water, help ease peptic ulcers, and is an excellent food for those trying to lose weight. You can add alfalfa to most dishes but it is ideal in salads and sandwiches. Try sprouting your own from seed for the freshest sprouts ever.

Almonds

Of all the different varieties of nuts, almonds have the most calcium and an even higher magnesium content, giving them a good ratio for an alkaline-forming food. They also contain essential fatty acids, vitamin A and some of the B-group, and the minerals phosphorus and potassium. Almonds help maintain strong bones and teeth because of their calcium and phosphorus content. They also lower harmful LDL cholesterol. One study found a 12 per cent decrease in LDL levels in people consuming 100mg of almonds a day for just nine weeks. Although all nuts have the ability to lower cholesterol levels, almonds have the most dramatic effect. Choose almonds in preference to other nuts as walnuts, hazelnuts, Brazils and most other nuts are acid forming.

Amaranth

Amaranth, which can be bought as flour or as grains, contains no gluten or yeast. It is an excellent source of calcium, magnesium and protein. The flour has a pleasant, nutty taste, making it the perfect choice for baking a variety of flat breads, bagels and cakes. Amaranth grains can be cooked and eaten like a cereal, or try adding it to soup for valuable protein. The grains can be gently toasted, which brings out their full flavour, and added to salads. Look out for it in health-food shops and experiment.

Apples

Apples are a versatile and potent package of natural goodness and deserve their reputation for keeping doctors away. They contain valuable bulk fibre in the form of indigestible cellulose, which is needed not only for regular bowel movements but also to carry toxins out of the body. Apples protect the heart, lower blood cholesterol and blood pressure, stabilize blood sugar and suppress the appetite. They contain high levels of calcium, magnesium, phosphorus, vitamin C and biotin. Be sure to eat the skin as it is high in pectin fibre, which is lost when apples are juiced. Apples also lose much of their nutritional value when dried. How many different varieties have you tried this week? Just the usual one? Be adventurous – there are dozens of varieties to choose from.

Apricots

The apricot is cherished in the Himalayan kingdom of Hunza (the land of Shangri-la in the novel and film *Lost Horizon*) as a source of health and exceptional longevity. Apricots are better eaten in their dried state as they have a much higher concentration of betacarotene than the raw fruit. Betacarotene converts to vitamin A in the body when needed. Apricots may protect against cancer, especially smoking-related cancers. They are extremely high on the list of alkaline-forming foods and we encourage you to eat them three times a week. They make an excellent mid-morning or afternoon snack. Add a few almonds for a more balanced snack. For some people, eating apricots alone may increase blood sugar levels too quickly, as they are also a concentrated source of sugar.

Artichokes

The artichoke is an edible thistle. Its therapeutic benefits may include lowering blood cholesterol levels and stimulating bile and urine, making it a natural diuretic. It is also a good aid to digestion. In 1969, French scientists were so successful in using artichoke extract for treating liver and kidney ailments that they took out a patent on it resulting in cynarin, a constituent of the artichoke, being formulated into a drug for lowering blood cholesterol. Russian scientists reported in 1970 that the edible parts of the artichoke had anti-inflammatory properties. Highly alkalizing, the artichoke makes an excellent starter to any meal.

Asparagus

There are many different varieties of asparagus, of which the green asparagus is the most common. Only the green tips are alkaline forming – the white stalk ends are acid forming so cut them off and discard them. Asparagus is not only much more freely available now but also more affordable. It is packed with nutrients, providing an excellent source of vitamin C and folic acid. Asparagus has been shown to protect against cataracts, macular degeneration (physical disturbance of the centre of the retina), atherosclerosis (hardened arteries), bruising and cancer, all due to its high content of antioxidants and alkalizing properties. Lightly cooked or raw in salads, asparagus adds variety and balance to any meal.

Aubergine (Eggplant)

With its satiny, purple skin, the aubergine is highly regarded in Nigeria, where it is used as a contraceptive, an anti-rheumatic agent and an anti-convulsant. In traditional Korean medicine, the dried plant, including the fruit, is consumed to treat lumbago, general pain, measles, stomach cancer and alcoholism, and is applied externally to cure rheumatism, gastritis and burns. Studies have shown that aubergines may also counteract artery damage.

Aubergines are an excellent source of betacarotene, which converts to vitamin A in the body when needed. The dish usually associated with aubergines is Greek moussaka, but they can be used in many ways to add colour and vibrancy to a meal, notwithstanding their alkaline effect on the body. Try the Vegetarian Moussaka in the recipe section (*page 126*) for a really substantial, but alkaline-forming, meal.

Avocados

Avocados are a wonderful health food, containing all the essential amino acids and the health-promoting essential fatty acids. They have excellent levels of magnesium, calcium and B-vitamins, and being alkaline forming too, what more could you want! Many people tend to avoid avocados because of their high calorie and fat content, but this is more than compensated for by the health-promoting properties of this versatile fruit. Fats are vital in the diet. An average avocado contains 30mg of total fats, but 23 per cent come from the health-promoting essential fatty acids of the omega-3 and omega-6 group. Avocados are classed as a protein food – so are excellent for vegetarians – and as one of the alkaline super foods in the pH Diet.

Bananas

Bananas have the most healing properties when allowed to ripen on the stem. Always choose the 'speckled' ones. Never choose green bananas as although they are alkaline forming they are difficult to digest and distinctly less healthy. Nutritionists and naturopaths use bananas to treat various conditions including hangovers, digestive disorders, heartburn, gas and painful menstruation. They can heal the stomach lining by triggering the release of a protective layer of mucus that rapidly seals off the surface, preventing stomach acids from causing damage. As such, bananas are excellent for preventing stomach ulcers. In cooking, they are a useful addition to warm savoury dishes. Although a good alkaline-forming food, bananas should be used sparingly as they are also a very 'heavy' food and can increase blood sugar levels quickly – eat no more than four a week.

Bean Sprouts

Bean sprouts, the sprouts of mung beans, are one of the most popular dietary staples in India and China, and are highly regarded in Ayurvedic and Chinese medicine. They have an alkalizing effect on the body and are lighter and easier to digest than most other beans. Bean sprouts are used widely for healing, especially in the restorative and cleansing Indian dishes *kichaki* and *dal*. Their effect on the body tends to be slightly cool but they can be warmed up with ginger, black pepper, cumin and mustard seeds without losing their alkalizing effect.

Sprouting your own mung beans is easy and inexpensive, but if you are short of time then you can buy them ready sprouted from most supermarkets. With careful planning, it is possible to eat a meal with a high percentage of alkaline-forming, and therefore health-maintaining properties, at a Chinese restaurant.

Beetroot

Beetroot sweetens, warms, moistens and adds mass to any meal and is an excellent way to introduce colour and interest to a dish. Used for uterine disorders, constipation and haemorrhoids, beetroot is also said to purify the blood and benefit the female reproductive system. Beetroot is rich in folic acid and a valuable source of vitamin A and potassium. It contains more magnesium than calcium, making it an excellent alkaline-forming food.

When was the last time you had beetroot in your diet? It tends to be something you either love or hate, but if you are not too keen or cannot remember the last time you tried some, reintroduce it into your diet to give colour, fibre and life to your food. Beetroot is usually bought ready cooked but can be obtained raw. Raw beetroot grated and added to salads and juices maintains even more of its alkaline properties.

Blackberries

Bursting with vitamin C, blackberry juice is also a good astringent antidote for diarrhoea. Not many people realize that blackberries are an excellent source of calcium, yielding 46mg to 1 cup. Surprisingly, blackberries are also a great source of fibre, giving almost 6g per cup compared to 0.5g for one banana. They are expensive to buy from supermarkets so get out in the countryside around September and pick your own. Frozen packets of mixed berries, including blackberries, strawberries, blackcurrants and loganberries, are available from supermarkets and are an excellent way to start the day. Mix some with bio yoghurt or grapefruit and a few almonds for balance. They are high in phytoestrogens, making them especially good for women over 40. Avoid blackberry drinks with added sugar or sweeteners as these are very acid forming.

Broccoli

Abundant in numerous strong antioxidants, including quercetin, betacarotene and vitamin C, broccoli has extremely high anti-cancer properties, particularly against lung, colon and breast cancers. Like other cruciferous vegetables, it speeds up the removal of oestrogen from the body, helping suppress soft-tissue cancers like breast cancer. Broccoli is also rich in cholesterol-reducing fibre and is an excellent source of chromium, which helps regulate insulin and blood-sugar levels. To get the most benefit from broccoli, steam lightly as cooking destroys some of the antioxidants and anti-oestrogenic agents. To get even more goodness, just blanch the broccoli florets and add to salads.

Brussels Sprouts

Brussels sprouts come from the cruciferous family so possess similar powers to broccoli and cabbage. They have anti-cancer actions and speed up the removal of oestrogen from the body, helping suppress soft-tissue cancers like breast cancer. They are also packed with various antioxidants. In one study, eating cabbage more than once a week cut the chances of men developing colon cancer by 66 per cent, and as little as two daily tablespoons of cooked cabbage protected against stomach cancer. Don't cook Brussels sprouts until mushy! Leaving them slightly crunchy will retain their alkaline-forming properties.

Buckwheat

Buckwheat is not related to any other wheat. It is a broadleaf crop belonging to the same plant family as rhubarb. It is particularly good for you because it contains rutin, which appears to have an uplifting effect on the spirits and is used by nutritionists to treat atherosclerosis (hardened arteries). It is also very good for people who bruise easily and have fine broken capillaries under the skin. Buckwheat can be used in a variety of different recipes, as a breakfast cereal or as an alternative to rice or pasta dishes.

Cabbage, Sauerkraut and Bok Choy

Cabbage is a true super food. Everyone should aim to eat it at least once a week. It is anti-bacterial and anti-viral, it heals stomach ulcers, manages oestrogen, prevents colon cancer and fights many other cancers, particularly stomach cancer. Cabbage is rich in vitamins A, B, C, K and E, potassium, sulphur and copper, plus a variety of antioxidants. Remember what your mother and grandmother told you about throwing away the cabbage water? So now you know – never throw away the water but make the gravy with it, or better still drink it.

For the most benefits, eat cabbage raw – as in coleslaw – or very lightly steamed. Chinese cabbage (bok choy) is said to have even more beneficial properties. Finnish researchers now advise us that fermented cabbage, otherwise known as sauerkraut, could be even healthier than raw or cooked cabbage.

Canola Oil

Canola oil is a monounsaturated fat, which is good for cooking as it can withstand high temperatures. It has a milder flavour than olive oil, so is more suitable for baking. Grown primarily in regions of western Canada, it has the best fatty-acid ratio of all oils. It has the lowest level of saturated fat (7 per cent), is relatively high in monounsaturated fat (61 per cent) and contains a moderate level of polyunsaturated fat (22 per cent). Together with being alkaline forming, it is a good oil to use. Remember to buy only cold-pressed oils.

Cantaloupe Melon

All melons, but cantaloupe melons in particular, are excellent sources of vitamin A due to their concentrated betacarotene content. Once inside the body, betacarotene can be converted into vitamin A by the liver. Cantaloupe is also a valuable source of vitamin C. In addition to its antioxidant activity, vitamin C is critical for good immune function. It stimulates white cells to fight infection, directly kills many bacteria and viruses, and supports vitamin E in the body. Cantaloupe also contains vitamins B_6, B_5 and B_1, potassium, dietary fibre and folic acid. The combination of these B-complex vitamins along with the fibre makes cantaloupe an exceptionally good fruit for supporting energy production. The fibre content helps ensure the sugars in the melon are delivered into the bloodstream gradually, keeping blood sugar on an even keel.

Melons, especially cantaloupe melons, can also be described as anticoagulants – they contain compounds that thin the blood. If the blood remains thin, it lessens the risk of heart attacks and strokes. Being an alkaline-forming food, cantaloupe is invaluable for people who eat out, as it is often found on menus. Ideally, melons should be eaten alone as they ferment very quickly in the stomach. If having melon as a starter, leave as much time after eating as possible before the next course.

Carob

The carob tree, a member of the legume family, is a slow-growing, evergreen tree originating in the Mediterranean. The seeds are used to make locust bean gum, which is used in food manufacture. The ground-up pod forms a high-protein powder and is an effective substitute for cocoa powder. Unlike cocoa, carob is free from the stimulants caffeine and theobromine, which can be addictive and cause allergies. It is also free from tyramine, which can trigger migraines and allergic reactions. It is often used to make 'healthy chocolate' and has become a popular chocolate substitute and a general sweetener.

Carob is 80 per cent protein, and contains vitamins A, B_1, B_2, B_3 and D. It is also high in calcium, phosphorus, potassium and magnesium and contains iron, manganese, barium and copper. It also has medicinal uses including the treatment of coughs and diarrhoea.

Carrots

Carrots have many therapeutic benefits including lowering blood cholesterol levels and preventing constipation. They are loaded with betacarotene, which has been shown to have a dramatic impact on lung cancer. A single carrot once a day appears to cut the risk of lung cancer at least in half, even among former heavy smokers! However, do not eat carrots in the mistaken assumption that they will allow you to continue smoking. Whereas carrots may cut your lung cancer risk in half, smoking boosts it 10 times as much.

Two-and-a-half raw, medium-sized carrots a day have lowered blood cholesterol by an average of 11 per cent. Raw carrots also increase the bulky weight of the stool by about 25 per cent, which helps keep the colon healthy.

To get the most benefit from carrots, eat them lightly steamed. Cooking releases carotenes, the active beneficial agents. You get two to five times *more* carotene from cooked carrots than from raw ones. Take care not to overcook them, however, as they will lose much of their health-promoting betacarotene and become a high-glycaemic food, raising blood-sugar levels quickly.

Cauliflower

Cauliflower is a cruciferous vegetable and a member of the cabbage family. Cruciferous vegetables contain a substance called indole-3-carbinol (I3C). This has been reported to affect the metabolism of oestrogen in a way that might protect against breast and other female cancers. Cauliflower is an excellent source of vitamin C and a good source of folic acid. Vitamin C is a powerful antioxidant and has anti-inflammatory actions. Diets high in fruit and/or vegetables are associated with a reduced risk of stroke. Look out for green cauliflower, which provides extra vitamin A and slightly more vitamin C.

Celery

Celery is a wonderful alkaline-forming food and excellent for clearing away acid wastes. These build up and clog the tissues of the body, causing conditions such as arthritis, diabetes and varicose veins. Celery is particularly high in organic sodium, the companion of potassium, which is essential for maintaining the correct consistency of body fluids. It also contains organic calcium and other minerals, which help to balance the nervous system. Celery contains eight different families of anti-cancer compounds, such as polyacetylenes, that detoxify carcinogens, especially nicotine.

Chestnuts

Chestnuts are listed separately to other nuts because they are differ-ent. With the exception of almonds and chestnuts, nuts are acid forming and usually have a high fat and calorie content. Chestnuts, on the other hand, are very low in calories and fat, and therefore just the thing for the nutrition-conscious person. Their high water con-tent – about 50 per cent – makes them more like vegetables.

Eat chestnuts either on their own as a snack, just boiled and peeled, or in a stuffing mix to balance out a Sunday roast. As a guide, five medium cooked and peeled nuts would provide 30g (1oz) of nuts which would yield approximately 54 calories, compared to 180 calo-ries for the same weight of hazelnuts, 170 calories for cashews and 166 calories for peanuts. So you can have three times as many chest-nuts for the same calories, with the added bonus of them being alka-line forming in the body.

Chicory

A relative of endive, chicory has curly, bitter-tasting leaves that can be eaten raw or cooked like greens. It can also be roasted and dried and is often added to coffee for aroma and flavour. Chicory is an excellent source of potassium, vitamin C, folic acid and vitamin A, and a good source of calcium, yielding 180 milligrams per 40 grams (1¹/₂ ounces). The calcium in green leafy vegetables like chicory is absorbed more easily than from dairy products, making it an excel-lent food for all adults. It is particularly valuable for adolescents and women who are pregnant or breastfeeding, as their calcium needs are greater. Eating plenty of vitamin C-rich vegetables helps to sup-port the structure of capillaries, making chicory a good choice for skin care too.

Chives

A member of the onion family, the chive is a fragrant herb with a mild onion flavour, but should be thought of as more than just a seasoning. Throughout history, chives have been used in natural healing. In the Middle Ages, they were even thought to ward off evil spirits. Today we know that chives are a very good source of vitamin C and contain good levels of minerals such as potassium, calcium, iron and folic acid to help promote a healthy body. Chives are helpful in easing stomach upsets, promoting an appetite, clearing stuffy noses and preventing bad breath. They make a tasty addition to omelettes, salads and many other dishes, and can be added freely to any acid-forming meal to redress the balance a little.

Courgettes (Zucchini)

Courgettes are probably the best known of the summer squashes. Related to both the melon and the cucumber, the entire vegetable, including its flesh, flowers, seeds and skin, is edible. The courgette and other members of its family have many health benefits, apart from being alkaline forming. Summer squash is an excellent source of manganese and magnesium. It also contains high levels of carotenoids (that convert to vitamin A in the body), the B-vitamins, potassium, folic acid and copper, and is a good source of fibre. The magnesium has been shown to be helpful in reducing the risk of heart attack and stroke. Together with the potassium, magnesium is helpful for reducing high blood pressure. Courgettes have also been found to help reduce symptoms of an enlarged prostate gland.

Cucumber

Cucumbers belong to the same family as pumpkin, courgettes (zucchini), watermelon and other types of squash. The flesh is primarily composed of water, making it naturally hydrating – a must for glowing skin. Cucumbers contain vitamin C and caffeic acid, which can soothe irritations and reduce swelling when applied to the skin.

Cucumber juice is often recommended as a source of silicon to improve the complexion and health of the skin. Its high water content eases various types of skin problems, including swelling under the eyes and sunburn.

The vegetable's hard skin is rich in fibre and contains a variety of beneficial minerals including silica, potassium and magnesium, so always eat the skin. Like other fragile vegetables, cucumbers may be waxed to protect them from bruising during shipping. Plant, insect, animal or petroleum-based waxes may be used. We recommend that you choose organically grown cucumbers, that are usually wax free.

Endive

Endive is a small, cylindrical head of pale, tightly packed leaves. A close relative of chicory, it has less nutritional value but is still alkaline forming. Endive is a good source of vitamin A, which is associated with a lower risk of cataracts. It makes an excellent addition to salads and combines well with most other fruits and vegetables. Try our easy-to-make Watercress and Endive Salad with Pears and Cheese (*page 131*) for a tasty, different and alkaline-forming dish.

Fennel

In the past, the only beneficial parts of fennel were thought to be the root and the seeds. However, the green leaves growing above the ground may also contain many beneficial properties and can be juiced. Fennel has the odour of liquorice and a sweet flavour. The juice on its own soothes the stomach nerves and is an effective remedy for flatulence and abdominal cramps. Fennel can be used successfully against colds for loosening and expelling mucus. When added to fish, it can help neutralize an acid-forming dish.

Figs

One of the most alkalizing of all the foods listed here, figs not only give us a great 'sweet' food, but they also have many therapeutic benefits. Figs help fight cancers and aid digestion. In juice form, they can kill bacteria and roundworms. With so many fruits and vegetables on our list, I guarantee you will be using figs in your future cooking just to get some sweetness back. Fresh figs, in particular, make an excellent snack – add some almonds for balance.

Flaxseed Oil

Also referred to as linseed, flaxseed is classed as a 'functional food' because it provides so many nutritional and health benefits. It contains ligans, which may play a role in preventing cancers of the breast, womb lining and prostate. Its fibre content helps lower cholesterol, regulate blood sugar levels and aid digestion, and its omega-3 fatty acids help lower the risk of cardiovascular disease and stroke. Flaxseed has a big role to play in the pH Diet. We recommend you eat some every day in the form of the '3&6 Mix' (see page 112). Flaxseed can also be introduced to the diet through omega-3 eggs, which are produced by hens fed on flaxseed-fortified rations.

Garlic

Garlic's smelly compound, allicin, was identified as an antibiotic in 1944. Tests found raw garlic to be even more powerful than penicillin. Garlic is well known for its ability to fight infections. It contains cancer-preventing chemicals, acts as an anti-coagulant (thins the blood), reduces blood pressure, cholesterol and triglyceride levels in the blood, stimulates the immune system, prevents and relieves chronic bronchitis and acts as an expectorant and decongestant.

Try to eat garlic every day. Chewing parsley stalks afterwards will help sweeten your breath, supposedly due to its chlorophyll content. Although only raw garlic will kill bacteria, cooked garlic retains its benefits of lowering blood cholesterol and will help to keep blood thin and perform as a decongestant. To get all the benefits, eat both raw and cooked garlic as often as possible.

Ginger

Reap the health benefits of fresh ginger in wonderful ginger tea (on its own or with lemon) or add it to soups and stews. Ginger prevents motion sickness, thins the blood, lowers blood cholesterol levels and may prevent some cancers. It is an excellent remedy for pregnancy sickness, and is warming to the body.

Goat's Cheese and Milk

Most people assume goat's milk will have the same strong, musky taste for which goat's cheese is famous. In fact, good quality goat's milk has a delicious, slightly sweet and sometimes slightly salty taste. Goat's milk is a very good source of calcium, vitamin B_2 and biotin, and a good source of low-cost, high-quality protein, vitamins B_5 and D and potassium. Biotin is involved in the metabolism of both sugar and fat, so drinking goat's milk can help boost energy production and promote a healthy skin and nervous system. It is an alkaline-forming food, unlike cow's milk, which makes it a healthy alternative.

Perhaps the greatest benefit of goat's milk is for those people who cannot tolerate cow's milk. Allergy to cow's milk has been found in many people with conditions such as recurrent ear infections, asthma, eczema and even rheumatoid arthritis. Replacing cow's milk with goat's milk may help to reduce some of the symptoms of these conditions. Like cow's milk, however, goat's milk contains the milk sugar lactose, and may produce adverse reactions in lactose-intolerant individuals.

Goat's cheese is superior to cow's cheese in that it contains much less saturated fat but is an excellent source of calcium. Never tasted it? Experiment and try the recipes in the next chapter.

Grapefruit

Grapefruit contains many health-promoting compounds. It is an excellent source of dietary fibre, vitamin C and vitamin A through betacarotene and a good source of potassium. Vitamin C helps to support the immune system and protects cells from free-radical damage. The rich pink and red colours of grapefruit are due to the phytochemical lycopene, which is thought to have anti-tumour activity. Other phytochemicals called limonoids also inhibit tumour formation. The pulp of citrus fruits like grapefruit contains glucarates, compounds which may help prevent breast cancer. Grapefruit also contains pectin, a soluble fibre that forms a gel-like substance in the intestines and can trap fats like cholesterol.

Grapefruit is an excellent breakfast fruit, makes great salsa and adds a tangy spark to green salads. Topped with berries, it makes an excellent alkaline start to the day.

Grapes

Grapes are well known for their health benefits. Said to be able to inactivate viruses, they thwart tooth decay and are rich in compounds that block cancer in animals. In her book, *The Grape Cure*, Johanna Brandt claims grapes cured her abdominal cancer. They are an excellent food for children and the elderly or for those recuperating from illness, as they are easy to eat and often sweet tasting. They are high in fruit sugar, making them an excellent accompaniment to cheese, as well as balancing out some of the acidity. They also go well with chicken and other savoury dishes.

Green Beans

Commonly referred to as string beans, although the string that was once their trademark can seldom be found in modern varieties, green beans are one of the few types of beans that are eaten fresh. They contain tiny seeds within their thin pods and are abundant in nutrients. Green beans have almost twice as much iron as spinach, which is an important mineral for menstruating women, who are more at risk of iron deficiency. In comparison to red meat, a well-known source of iron, green beans provide much more iron for a lot fewer calories and are totally fat free.

Green beans are an excellent source of protein, fibre, vitamin K and carotene (which converts to vitamin A in the body) and a very good source of vitamins C, B_2, B_3, B_5, B_6 and folic acid. They are also rich in the minerals potassium, iron, manganese and magnesium, not to mention the calcium, copper and zinc.

The vitamin K provided by green beans is important for maintaining strong bones. Magnesium and potassium work together to help lower high blood pressure, while folic acid and vitamin B_6 are needed to convert a potentially dangerous molecule called homocysteine into other, less dangerous molecules. With all these benefits, you will want to try our recipe for Green Beans and Almonds on page 133.

Honey

Honey is 'manufactured' in one of the world's most efficient factories – the beehive. Bees may travel up to 55,000 miles and visit more than two million flowers to gather enough nectar to make just a pound of honey. There is no doubt that honey has healing properties, and its use can be traced back to Egyptian times. However, its therapeutic activity differs according to the types of flowers the nectar has been gathered from. Honey can be made from clover, eucalyptus and lavender, to name just three.

Honey is composed primarily of fructose, glucose, maltose and water. It also contains other sugars as well as small amounts of trace enzymes, minerals, vitamins and amino acids. Honey can destroy harmful bacteria that can cause digestive upset and promote beneficial bacteria in the gut. Anti-microbial and anti-bacterial, honey stimulates the immune system and is used to treat stomach ulcers and sore throats. Used topically, honey can be applied to burns, varicose veins, skin ulcers, wounds, acne, eczema and boils.

Certain honeys are particularly high in anti-bacterial properties, the most notably active being Manuka (or tea tree) honey from New Zealand. Excellent results were found in recent clinical trials held in New Zealand when active Manuka honey was used on previously unresponsive skin ulcers and wounds. Manuka honey can be effective against the *H. pylori* bacteria that are implicated in peptic ulcers. Active Manuka honey is also effective in treating bacterial gastroenteritis in infants, and there is evidence that some strains of honey are prebiotic (stimulating the activity of the body's existing good bacteria). Honey is not the only alkaline super food with these gut flora-boosting properties – others include chicory, artichokes, garlic, onions, leeks, asparagus, peaches and bananas.

Kale

How often have you seen kale in the supermarket and walked right past it? Yet this unsung health hero is an amazing super food, packed with vitamin C. It is a good source of calcium and magnesium and contains small amounts of *all* eight essential amino acids.

The dark-green leafy vegetables we hear so much about are the most protective, and kale is top of the list. Kale is loaded with different types of betacarotene, which are then converted to vitamin A in the body. Kale has twice as much betacarotene as spinach, and some scientists believe it to be a potent anti-cancer agent. It is also an excellent source of chlorophyll, another well-known anti-cancer agent. As a member of the cruciferous family, kale is also thought to lower rates of bowel, prostate and bladder cancer.

Kale is wonderful eaten raw in salads. As with other vegetables rich in betacarotene, however, light cooking enhances the effect of this nutrient. Include both raw and lightly cooked kale in your diet to get the maximum benefit.

Kelp and other Sea Vegetables

The sea yields several substances that are excellent at removing harmful acid wastes from the body. All seaweed – including kelp, nori and dulse – is rich in trace elements and minerals, which aid the body's metabolic function. Seaweed is also an extremely rich source of calcium and iodine. Coarse hair and dry skin may be signs of low iodine levels in the diet. Another symptom is a lack of energy, so the more sea plants you can incorporate into your diet the better. Seaweed fibre can bind and remove waste materials from the body, and is a good detoxifier for women with cellulite.

Spirulina is fresh-water algae with the richest iron content known in nature. It also contains more vitamin E than wheatgerm and up to six times more vitamin B_{12} than raw beef. It is a good supplement to take if you suffer from cellulite. You can buy it in tablet form or as a powder, which you can make into a hot drink or mix with light vegetable soup.

Kiwi Fruit

Kiwi fruit is native to China, where is it called *yang tao*. The Americans named it after the native bird of New Zealand, whose brown, fuzzy coat resembles the skin of this unique fruit. Packed with more vitamin C than an equivalent amount of orange, the bright green flesh of the kiwi fruit speckled with tiny black seeds adds a dramatic, tropical flourish to any meal.

Kiwi contains numerous health-promoting nutrients. Research has found that it has the ability to protect DNA in the nucleus of human cells from oxygen-related damage. Kiwi fruit is a good source of two of the most important fat-soluble antioxidants, vitamin E and betacarotene. This combination of both fat- and water-soluble antioxidants makes kiwi able to provide protection from free radicals on all fronts.

Kiwi fruit is also an excellent source of dietary fibre. Diets containing plenty of fibre can reduce high cholesterol levels, which in turn may lower the risk of heart disease and heart attacks. Fibre is also good for binding and removing toxins from the colon, which is helpful for preventing colon cancer. In addition, fibre-rich foods like the kiwi are good for keeping blood sugar levels under control. Add kiwi fruit to tossed green salads, fruit salads or eat alone. Its alkalizing properties will assist with balancing any meal.

Leeks

Leeks are part of the allium family, which includes garlic, onions, shallots and spring onions (scallions), to which they bear a resemblance. They contain many of the same beneficial compounds found in garlic and onions. Leeks are a good source of vitamins B$_6$ and C, folic acid, manganese, iron and dietary fibre. This particular combination of nutrients makes leeks particularly helpful in stabilizing blood sugar, since they not only slow the absorption of sugars from the intestines but also help metabolize them in the body.

Regular consumption of allium vegetables – as little as twice a week – is associated with a greatly reduced risk of colon cancer. This is because several of the compounds found in these foods are able to protect colon cells from cancer-causing toxins, while also stopping the growth and spread of any cancer cells that do develop. With a more delicate and sweeter flavour than onions, leeks add a subtle touch to recipes.

Lemons and Limes

Lemons and limes gained their fame for their ability to prevent scurvy, a disease caused by a severe deficiency of vitamin C. Scurvy was the curse of British seamen who went for months without fresh fruits and vegetables. A little over a tablespoon of lemon juice daily prevents scurvy because of the vitamin C content.

Scurvy is rare today, but as vitamin C cannot be made or stored in the body, it must be obtained from the diet on a daily basis. Vitamin C travels through the body, neutralizing any free radicals with which it comes into contact. Free radicals can damage healthy cells and their membranes, and also cause inflammation in the body. This is one of the reasons why vitamin C is helpful in reducing some of the symptoms of osteoarthritis and rheumatoid arthritis. Give your body a great start to the day with a glass of water with a tablespoon of alkaline-forming lemon or lime juice.

Lettuce

There are hundreds of varieties of lettuce including iceberg, romaine, radicchio, green and red leaf, escarole, endive, watercress and Boston. Most types of lettuce provide a small amount of folic acid, vitamins A and C, potassium and fibre. The darker green leaves provide more nutrition than the light green or yellow varieties. Romaine lettuce has good amounts of vitamins compared with other varieties. A cup of romaine lettuce would provide 13.4mg of vitamin C and 76mg of folic acid; iceberg lettuce would provide only 5mg and 0.02mg respectively. Try adding lettuce to salads and sandwiches for its texture and nutrients. It is good for balancing any acid-forming foods.

Mangos

The mango, which is related to pistachios and cashews, has been known for some 6,000 years and is native to India, where the trees grow to about 15 metres (50 feet). Mangos are perfect eaten raw as they are exquisitely sweet, but they should always be peeled as the skins are very acidic. They are an excellent source of vitamin A in the form of betacarotene, and vitamin C. Vitamin A, like mangos themselves, is soothing and regenerative for the digestive system and throat. It also encourages clear, healthy skin. Mangos have been used in the Ayurvedic system of healing for centuries for these purposes.

An excellent fresh mango salsa can be made from the unripe fruit and used to accompany and balance acid-forming protein meals. A chicken or meat curry, for example, would be acid forming, but serving it with fresh mango salsa would make it less so. (*See recipe on page 135.*)

Millet

Millet is a wonderful grain. It can be creamy like mashed potatoes or fluffy like rice, depending on the cooking method. Since millet contains no gluten, it is a wonderful alternative for people who are gluten-sensitive. Millet is tiny and round, and the most widely available form is the pearled, hulled type. You may also be able to find traditional couscous, which is made from cracked millet.

As well as being one of the few alkalizing grains, millet is also a good source of some important nutrients, including vitamins B_1 and B_3, phosphorus and magnesium. The body cannot produce energy without adequate supplies of vitamin B_1, and it also plays a key role in supporting the nervous system. The good source of phosphorus provided by millet is important in the structure of every cell in the body. Cooked millet can be served as a breakfast porridge to which you can add your favourite nuts and seeds for balance.

Mushrooms

Common button mushrooms are an excellent source of chromium. However, the mushrooms with the most health-giving properties are the speciality mushrooms: shiitake, maitake and reishi. These mushrooms boost the immune system, combat viruses and bacteria, protect the body from cancer, increase energy, ward off hunger, extend longevity, promote overall vitality and virility, lower blood cholesterol levels, balance blood pressure and inhibit platelet aggregation (reducing the likelihood of heart disease). Care should be taken with mushrooms if you suspect you have a candida (yeast) infection as mushrooms, being fungi, are on the no-go list – alkaline forming or not!

With their ancient heritage, these simple, basic super foods appear to have immense potential for human health. They may even hold the key to some of the world's most serious diseases. If you've never tried speciality mushrooms, spoil yourself, and remember that quality, not quantity, is what matters.

Okra

Okra is excellent stewed with tomatoes and onions, which makes a good base for many meals, or can be gently sautéed in olive oil as an alternative vegetable. When boiled, it gives off a viscous substance that can add smoothness and thickness to soups and stews. Described as cool, diuretic, soothing and softening, okra has an anti-inflammatory action that can alleviate the symptoms of cystitis, sore throats, fevers, bronchitis and irritable bowel syndrome. Okra is an excellent source of vitamin C, folic acid and vitamin A in the form of betacarotene. It is also a good source of magnesium. Next time you see it, put it in your shopping basket and experiment.

Olive Oil and Olives

Monounsaturated fats such as olive oil should make up 10 per cent of a balanced diet. Olive oil is suitable for cooking at low, medium and high temperatures. Rich in flavour, it is used frequently in marinades, sauces and salad dressings. The nutritional content, storage life and quality of olive oil depend upon the processing methods used. For the best quality oil, choose only cold-pressed virgin olive oil. Virgin olive oils are produced from the first pressing of the olives, and are unrefined. As a result, these oils are more nutritious. Select oils in light-resistant plastic containers or dark brown or green glass containers. Olive oil is manufactured all over the world. Experiment and buy different oils to compare their flavours.

Olives add a wonderful taste to salads. There are hundreds of different varieties to choose from. Olives are a good source of iron, dietary fibre and calcium.

Monounsaturated fats, such as olive, peanut and canola oils, have a protective effect on cells. Olive oil also contains the antioxidant vitamin E, which can lower the risk of cell damage and inflammation. Olives also contain a variety of beneficial active compounds called polyphenols that appear to have significant anti-inflammatory properties. This combination of nutrients may help reduce the severity of asthma, osteoarthritis and rheumatoid arthritis, three conditions in which most of the damage is caused by high levels of free radicals. The vitamin E in olives may even help to reduce the frequency and/or intensity of hot flushes in women going through the menopause.

Onions

The onion is not only packed with therapeutic compounds similar to those found in garlic but also has unique properties of its own. It was Louis Pasteur, back in the mid-1800s, who first put onions to the test and declared them antibacterial. The onion has been cultivated for over 6,000 years. Its therapeutic benefits – whether eaten raw, roasted or cooked into syrup – have been valued for centuries.

Like garlic, onions are members of the allium family. Both are rich in powerful sulphur-containing compounds that are responsible for their pungent odours and for many of their health-promoting effects. Onions are also rich in chromium, a trace mineral that helps cells respond to insulin, plus vitamin C and numerous flavonoids, particularly quercetin.

Regular consumption of onions has, like garlic, been shown to lower high cholesterol levels and high blood pressure, thus helping to prevent atherosclerosis (hardened arteries) and reduce the risk of heart disease. These beneficial effects are probably due to the sulphur compounds in onions, as well as the chromium. Vitamin B_6 helps prevent heart disease by lowering high homocysteine levels.

Raw onions are best for raising beneficial cholesterol. You will have to eat approximately half a medium-sized onion every day for two months to get noticeable beneficial results from a cholesterol test. For other heart-protective effects, lightly cooked onions are as effective as raw. But don't forget – the more raw a food, the more alkaline it is.

Eating onions at least two or three times a week is associated with a significantly reduced risk of developing colon cancer. Quercetin has been shown to halt the growth of tumours in animals and to protect colon cells from the damaging effects of certain cancer-causing substances. Cooking meats with onions may help reduce the amount of carcinogens produced when meat is cooked in certain ways. Being so versatile, onions play a big role in the pH Diet to perk up salads, as a guacamole salsa dip, and sautéed and added to soups for flavour.

Oranges

Oranges are one of the most popular fruits around the world and make the perfect snack. They can transform an ordinary meal into a special one when used as an accompaniment or recipe ingredient. Although packaged orange juice has some beneficial qualities, it does not compare with the real fruit.

Not only are oranges an excellent source of vitamin C, they are also high in dietary fibre, which has been shown to reduce high cholesterol levels and thus help prevent atherosclerosis (hardened arteries). Fibre also helps keep blood sugar levels under control, which may help explain why the natural fruit sugar in oranges – fructose – does not raise blood sugar levels too dramatically. The fibre in oranges can also protect the colon. It attaches to cancer-causing chemicals, keeping them away from cells in the colon and taking them out of the body in the faeces. For those suffering from irritable bowel syndrome, the fibre in oranges may be helpful for reducing constipation and diarrhoea.

Oranges contain large quantities of phytonutrients. One that has been singled out as particularly beneficial is a flavanone called herperidin. This has been shown to lower high blood pressure as well as cholesterol in animal studies, and to have strong anti-inflammatory properties. Importantly, most of this is found in the peel and inner white pulp of the orange, rather than in its liquid orange centre, so this beneficial compound is often removed by the processing of oranges into juice. Eat as much of the white pulp as you can.

Papaya

Papaya is deliciously sweet with musky undertones and a soft, butter-like consistency. It is no wonder that Christopher Columbus reputedly called it the 'fruit of the angels'. Papaya is a rich source of antioxidant nutrients such as carotenes, vitamin C and flavonoids; the B-vitamins folic acid and pantothenic acid; the minerals potassium and magnesium; and fibre. Together these nutrients promote the health of the cardiovascular system and provide protection against colon cancer. The fibre may also help with the symptoms of irritable bowel syndrome. In addition, papaya contains the digestive enzyme papain, which is used to treat sports injuries, other causes of trauma, and allergies.

Most people discard the big black seeds but they are actually edible and have a delightful, peppery flavour. They can be chewed whole or blended into a creamy salad dressing to give it a peppery taste.

Parsley

Parsley is a versatile herb that makes a perfect accompaniment to any meal. With its high concentration of antioxidants, it can help detoxify carcinogens and neutralize certain cancer-causing substances in tobacco smoke. It is also a natural diuretic and good for the adrenal and thyroid glands. It helps maintain a healthy genito-urinary system and is used to treat kidney problems and eye disorders.

Parsley juice is very potent so use in small amounts with carrots and/or celery.

Parsnips

Parsnips make a useful addition to the pH Diet because as well as being alkaline forming, they also add sweetness to a meal and are very versatile. Parsnips can be baked, sautéed, steamed, boiled and mashed like potatoes. They are an excellent source of vitamin C and folic acid and a good source of potassium. Potassium reduces the amount of calcium excreted in the urine, and people with high amounts of potassium in their diet appear to be at low risk of forming kidney stones. The best way to increase potassium is to eat fruits and vegetables, as the levels of potassium in food are much higher than the small amounts found in supplements.

Peaches

Peaches are related to apricots, almonds, cherries and plums, and are good sources of vitamin C. They are also quite a 'heavy' fruit and can sometimes be difficult to digest. Only choose and eat fresh, ripe peaches. For easy peeling, blanch in boiling water for a few seconds, then plunge into cold water until cool enough to handle when the skin will slip off. Add them to yoghurts or sorbets, or spice them up to make an excellent side dish with any acid-forming meal to adjust the balance in your favour. Try the recipe for Poached Peaches (page 116) for a delightful summer dessert.

Pears

Just-ripe pears are a great snack, but you can eat this versatile fruit in salads, starters, desserts and main meals with poultry or meat. They make the perfect accompaniment to cheese and, because of their alkaline-forming nature, will help neutralize the acidity in a meal. Pears are low in calories, are an excellent source of fibre and provide small amounts of phosphorous, potassium and vitamin C. There are countless varieties of pears grown around the world and many are available in this country. Be adventurous and buy some you have never tried before.

Peas

The pea's greatest claim to fame in folk medicine is as an anti-fertility agent! Scientific studies have now backed this up, showing that peas do, in fact, contain well-known anti-fertility substances. After identifying the compound in peas, Indian researchers concentrated it, put it into capsules and gave it to women and men in controlled trials. Unbelievably, their pregnancy rate went down by 50–60 per cent. When men took the pea capsule, their sperm count dropped by half. Peas contain oestrogenic chemicals considered by experts to have some contraceptive activity. So if you are trying to conceive, keep off the peas. Peas also lower levels of detrimental LDL blood cholesterol because they are rich in soluble fibre.

Peppers

Sweet (bell) peppers belong to the nightshade family, which also includes aubergines (eggplants), tomatoes and white potatoes. Brightly coloured peppers – whether green, red, orange or yellow – are rich sources of some of the best nutrients available. To start, peppers are excellent sources of the powerful antioxidants vitamin C and betacarotene. These work together to neutralize free radicals, which can travel through the body causing huge amounts of damage to cells (*see page 43*).

Peppers are also an excellent source of vitamins K and B_6, and very good sources of dietary fibre, manganese, vitamin B_1 and folic acid. This versatile vegetable is widely used in the ph Diet.

Pineapples

The exceptional juiciness and vibrant flavour of pineapples make them one of the most popular tropical fruits. Pineapples have many health benefits. Fresh pineapple is rich in bromelain, a protein-digesting enzyme that not only aids digestion but can also effectively reduce inflammation and swelling. Bromelain has been shown to reduce swelling in inflammatory conditions such as acute sinusitis, sore throat, arthritis and gout, and speed recovery from injuries and surgery. Eat pineapple after meals that contain protein to assist digestion. To maximize bromelain's anti-inflammatory effects, eat it alone between meals.

Pineapple is an excellent source of manganese, which is needed by a number of enzymes involved in energy production and antioxidant defences. Pineapples are also a good source of two B-vitamins important in energy production – thiamin and riboflavin – and are an excellent source of vitamin C. Vitamin C is the body's primary water-soluble antioxidant, defending all aqueous areas of the body against free radicals that attach and damage normal cells (*see page 43*).

Pineapple can be left at room temperature for one or two days before serving. While this process will not make the fruit any sweeter, it will help it become softer and juicier. As they are very perishable, however, you should watch them closely during this period to ensure they do not spoil.

Potatoes

The potato belongs to the nightshade family, whose other members include tomatoes, aubergines (eggplants) and peppers. A baked potato (without the butter, sour cream, melted cheese and bacon bits!) is an exceptionally healthy, low-calorie, high-fibre food that offers significant protection against cardiovascular disease and cancer.

Potatoes are a very good source of vitamin B_6 and a good source of vitamins C, B_3 (niacin) and B_5 (pantothenic acid), dietary fibre and the minerals copper, potassium, iron and magnesium. Vitamin B_6 is involved in more than 100 enzymatic reactions. Enzymes are proteins that help chemical reactions take place, so vitamin B_6 is active virtually everywhere in the body. Many of the building blocks of protein – amino acids – require B_6 for their synthesis, as do the nucleic acids used in the creation of our DNA. Vitamin B_6 is therefore essential for the formation of virtually all new cells in the body. It is also necessary for the breakdown of glycogen, the form in which sugar is stored in our muscle cells and liver, making this vitamin a key player in athletic performance, endurance and energy.

The potato skin is a concentrated source of dietary fibre, so to get the most nutritional value from this vegetable, don't peel it – eat both the flesh and the skin. Just scrub the potato under cold running water before cooking. If you must peel, do so as thinly as possible to retain the nutrients that lie just below the skin.

Don't keep potatoes in the refrigerator because their starch content will turn to sugar, giving them an undesirable taste. Also, keep potatoes away from onions, as the gases they each emit will cause the other to decay. Store your potatoes in a paper bag – only use a plastic bag if it is perforated to allow moisture to escape.

Pumpkin

Pumpkins are 90 per cent water. They are packed with vitamin C, vitamin A in the form of betacarotene, and are an exceptional source of potassium. Pumpkins are also low in fat and calories and high in fibre.

A severely under-used food source, the pumpkin is sweet and heavy. Warmed and spiced, it makes substantial soup. Added to other vegetables, it makes a vital alkaline contribution to any meal.

Quinoa

Most commonly considered a grain, quinoa is actually a relative of leafy green vegetables like spinach and Swiss chard. It is a recently rediscovered ancient 'grain', once considered 'the gold of the Aztecs'. Quinoa is a protein-rich seed with a fluffy, creamy, slightly crunchy texture and a somewhat nutty flavour when cooked. It will play a starring role in your new alkaline-forming regime as it can be substituted for other grains, the majority of which are acid forming.

Quinoa is high in protein, but the protein it supplies is also complete – which means that it includes all eight essential amino acids. This makes it a good choice for vegans who may be concerned about adequate protein intake. Quinoa is especially well endowed with the amino acid lysine, which is essential for tissue growth and repair. In addition, quinoa features a host of other health-building nutrients. Because quinoa is a very good source of magnesium, zinc, dietary fibre and vitamins B_2 and E, this 'grain' may be especially valuable for people who suffer from migraine headaches, diabetes and atherosclerosis (hardened arteries). Quinoa also has a low gluten content, and is one of the least allergenic 'grains' there is.

Radishes

The radish is actually the root of a plant related to mustard. Its flavour varies from mild to peppery and pungent, depending on the variety – but all varieties are highly alkalizing, and therefore assist in balancing any acid-forming foods in a meal. Radishes come in a number of varieties, ranging in colour from red to purple to white, and in shape from small and round to long and oval. The most common radish is the oval, red-skinned variety, about the size of a cherry tomato. Look out for the Daikon radish, which is a long, white cylindrical variety that may weigh a pound (450g) or more, and is used primarily in Indian and Japanese cooking. All radishes are an excellent source of vitamin C, but the Daikon variety is additionally high in potassium, folate and magnesium. The high vitamin C content is excellent for anybody who bruises easily and can effectively lower homocysteine levels. So when next preparing a salad – don't forget the radish.

Raisins

Raisins are grapes that have been dehydrated using the heat of the sun or an oven. Among the most popular types are sultana, malaga, monukka and muscat. Raisins have been researched primarily for their unique phenol content, but these delicious dried grapes are also one of the top sources of the trace mineral boron. Critical to our health, particularly in relation to the bones, boron is required to convert oestrogen and vitamin D to their most active forms. Studies have shown that boron provides protection against osteoporosis and reproduces many of the positive effects of hormone-replacement therapy in post-menopausal women.

Sorrel

Although sorrel has an acidic taste that comes from the oxalic acid contained in its leaves, it is in fact alkaline forming in the body. The leaves have a slightly lemony flavour, which is perfect for fish, soups, white sauces, eggs, poultry and white meats. It also makes a good accompaniment to goat's cheese. Sorrel is rich in magnesium and potassium, and as such adds to your mineral reserves very nicely. It is also a good source of vitamin C.

Soya Beans, Cheese and Milk

In spite of its name, the soya bean is part of the pea family and is a legume. Soya beans are sometimes referred to as the 'new vegetable', and are the basis of all soya products. They are high in fibre, calcium, iron and protein, contain no cholesterol and are low in calories. Soya bean products include tofu, tempeh and miso, a thick high-protein paste, and are all excellent for creative cooking. Soya is rich in phytoestrogens. Research has shown that communities with diets high in phytoestrogen-rich foods enjoy significant health benefits.

While the soya bean packs a nutritional punch, the way soya beans are processed can alter the potentially beneficial phytochemicals. Whichever soya product you buy, ensure that the beans have not been genetically modified.

Soya Cheese
Soya cheese is a good substitute for most English and European cheeses, which are usually high in calories and saturated fats. Although there are other cheeses lower in saturated fat, like Edam, none compare to soya cheese, which is low in calories, low in saturated fats and high in health-promoting phytoestrogens. Although quite an acquired taste, it has the huge benefit of being alkaline

forming in the body, and as such makes a valuable contribution to meals and snacks on the ph Diet.

Soya Milk
Cow's milk does not suit humans well. It is a very acid-forming sub-stance, and the balance between calcium and magnesium is not cor-rect for human beings. Soya milk can be a good alternative to cow's milk as it will not only provide an invaluable source of calcium, but it is also low in fat and is alkaline forming. Like most high-protein foods, it promotes building and repairing of the body tissues.

Although soya milk used to taste awful, there are some great-tasting ones now on the market. Try spicing up your soya milk with a little cinnamon, cardamom, nutmeg or ginger and black pepper.

Spinach

Think of spinach and it is difficult not bring to mind Popeye, the cartoon sailor, who ate spinach to give him strength to fight off his wife's admirers! Spinach has been hailed as the king of vegetables by many leading scientists. It has extremely high concentrations of carotenoids, including betacarotene, which has been shown to pre-vent the promotion of certain cancers. Spinach also has high amounts of chlorophyll, another potential cancer-preventative food.

Spinach is especially beneficial for smokers or those trying to reduce smoking. Studies have shown that spinach can neutralize some of the damage to lung cells by blocking the promotion or progression of cancer. As an alkaline-forming, health-promoting food, spinach should be eaten at least once a week. However, cooked spinach also contains a compound called oxalid acid, which de-creases the gut's ability to absorb some minerals such as iron. So eat spinach leaves raw in a salad, or very slightly steamed, to avoid this drawback.

Strawberries

An excellent source of vitamin C and betacarotene, strawberries also supply good amounts of calcium and magnesium, and even contain traces of all eight essential amino acids. Studies have shown that strawberries can destroy viruses, such as polio and herpes simplex. Strawberries are also linked to lower cancer deaths.

With its seeds on the outside, the strawberry is indeed different. Its 99-per-cent water content makes it an excellent anti-diuretic. Strawberries are colourful, exciting, taste great and are an invaluable food when eating out as you nearly always find strawberries on the menu. Forget the cream. Try them 'sweet and sour' with freshly ground black pepper for the added advantage of the zinc in the black pepper – and to be different!

Sweetcorn

Sometimes referred to by its traditional name of maize, sweetcorn is a very good source of vitamins B_1, B_3, B_5, folic acid and vitamin C. It is also an excellent source of dietary fibre, and a good source of the minerals phosphorous and magnesium. Foods high in magnesium are often alkaline forming, and sweetcorn is no exception. High-magnesium foods improve the flow of blood, oxygen and nutrients throughout the body. Due, in part, to magnesium's relaxant effects, sweetcorn has also been shown to reduce the severity of conditions like asthma and migraine, lower high blood pressure and reduce the risk of atherosclerosis (hardened arteries) and heart disease. Traditional corn on the cob makes a filling starter. Adding sweetcorn to soup enhances the soup's bulk and nutritional profile.

Sweet Potato

There is often much confusion between sweet potatoes and yams. The moist-fleshed, orange-coloured root vegetable that is often called a yam is actually a sweet potato. The sweet potato is an excellent source of vitamin A, in the form of betacarotene, a very good source of vitamins C and B_6 and a good source of manganese, copper, biotin, pantothenic acid (vitamin B_5), vitamin B_2 and dietary fibre. With all these nutrients, sweet potatoes have antioxidant and anti-inflammatory healing properties.

Interestingly, the sweet potato is also classified as a 'diabetic' food. It has been given this label following recent animal studies in which sweet potato helped stabilize blood sugar levels and lower insulin resistance.

Sweet potatoes are grouped into two different categories, depending upon the texture they have when cooked. Some are firm, dry and mealy, while others are soft and moist. Both types are starchy and sweet with different varieties having unique tastes. Experiment with both types. If you have a sweet tooth, sweet potatoes will become very important in your diet, as they are one of the few truly sweet foods that are alkaline forming in the body.

Sweet potatoes should be stored in a cool, dark and well-ventilated place, where they will keep fresh for up to 10 days. Store them loose, not in a plastic bag. Keep them away from sunlight or temperatures above 60°F (15°C) as this will cause them to sprout or ferment. Uncooked sweet potatoes should not be kept in the refrigerator. If you purchase organically grown sweet potatoes, you can eat the entire tuber, flesh and skin. Conventionally grown ones should be peeled before eating since the skin is sometimes treated with dye or wax. If preparing the sweet potato whole, just peel it after cooking.

Tamarind

The rind of a long fruit filled with large, hard seeds, tamarind gives the distinctive sour taste of Indian food. It is good with eggs or fish and excellent in curries and pickles, the seed pods adding a sour, prune-like flavour. Do not let that description put you off; combined with other ingredients, it adds a unique and delicious taste. Like many Indian foods, it is good for the digestion, and its high vitamin C content makes it a natural mild laxative. Tamarind is rich in vitamins and said to be a tonic for the kidneys and liver.

Tangerines

A cross between the mandarin orange and the bitter orange, the tangerine is named after the main mandarin shipping port of Tangier in Morocco. Tangerines are excellent sources of vitamins A and C, and because they are small and easy to peel, they may be more suitable for children and the elderly. As far as the ph Diet is concerned, tangerines make the perfect dessert on Level 3 as they are so quick and easy to prepare and eat. When added to fruit juices, they make a pleasant change from orange juice. The citric acid found in tangerines and all citrus fruits may also prevent kidney stones forming.

Tofu

Although once found only in Asian food markets, tofu is now widely available in supermarkets. This bland food can miraculously take on the flavours of its surrounding ingredients, making it a versatile as well as nutritious part of a healthy diet. Tofu is a protein-rich food made from the curds of soya milk. Off-white in colour, it is usually sold in rectangular blocks. Tofu is a staple in the cuisines of many

Asian countries. Tofu is its Japanese name, while in China it is known as *doufa*.

Research has shown that a regular intake of soya protein can help lower total cholesterol levels by up to 30 per cent, lower LDL (harmful cholesterol) levels by as much as 30–40 per cent, lower triglyceride levels, reduce the tendency of platelets to form blood clots, and possibly even raise levels of HDL (beneficial cholesterol). Soya has also been shown to be helpful in alleviating the symptoms associated with menopause. Soya foods, like tofu, contain phytoestrogens. In a woman's body, these compounds can act like very weak oestrogens. At the start of the menopause, when a woman's oestrogen levels fluctuate, the phytoestrogens in soya can help her maintain balance. They block out oestrogen when levels rise excessively high and compensate when levels are low. When a woman's production of natural oestrogen drops at menopause, the isoflavones in soya may provide just enough oestrogenic activity to prevent or reduce uncomfortable symptoms like hot flushes.

Most types of tofu are enriched with calcium, which can help prevent the accelerated bone loss for which women are at risk during menopause. If you are looking for tofu with a high calcium content, choose products that specifically state 'calcium enriched' on the label. Tofu is an excellent source of tryptophan, calcium, manganese, iron, selenium and zinc. In 115 grams (4 ounces) of tofu, there is 9.2 grams of protein, virtually no saturated fat (less than 1 gram), and only 86 calories. The firmer types of tofu are usually the highest in fat, and the softest – often called silken – are the lowest.

Tomatoes

Tomatoes are a major source of lycopene, an extremely potent antioxidant and anti-cancer agent. In particular, they are linked to lower rates of pancreatic and cervical cancer.

Watercress

Watercress is exceptionally high in sulphur, which is needed for strong nails and hair. It is a magnificent cleanser and is helpful, in combination with other juices, for treating anaemia, haemorrhoids (piles) and emphysema. It can be extremely bitter in juices so use just a little mixed with carrot, apple or other vegetables. Watercress also makes a very good soup but loses some of its beneficial properties when cooked.

Watermelon

The watermelon is related to the cantaloupe melon, squash and pumpkin, and other plants that grow on vines on the ground. It is an excellent source of vitamin C and a very good source of vitamin A, notably through its concentration of betacarotene. These powerful antioxidants travel through the body neutralizing acidity and free radicals. Watermelon is rich in the B-vitamins necessary for energy production. It can be described as a nutrient-dense food because it has a higher water content and lower calorie content than many other fruits, delivering more nutrients per calorie – an outstanding health benefit.

Wheatgrass

Wheatgrass is grown from a special strain of wheat that produces high concentrations of chlorophyll, active enzymes, vitamins, minerals and amino acids. Chlorophyll comprises 70 per cent of wheatgrass. Often referred to as 'the blood of plant life', chlorophyll closely resembles the molecules of human red blood cells. With so many similarities in structure, it is absorbed quickly through our digestive system and begins rebuilding our red blood cells and bloodstream. Just 30 grams (1 ounce) of wheatgrass juice is equivalent in vitamins, minerals and amino acids to that found in 1 kilogram (2¹/₂ pounds) of green leafy vegetables!

Wheatgrass purifies the blood, helps cleanse the liver and neutralizes toxins and carcinogens in the body. Fresh wheatgrass may be hard to come by unless you live in a city where there may be juice bars, or grow your own, but there are many outlets for freeze-dried wheatgrass (*see Resources*).

Wild Rice

With its many varieties, rice is probably the most versatile grain of all. Wild rice, however, is not really rice but a long-grain marsh grass native to the northern Great Lakes. A cholesterol-free, low-fat food, wild rice is high in complex carbohydrates, making it an excellent source of energy. Wild rice is also a good source of fibre and vitamin B_6, and contains no sodium. It has a nutty flavour and chewy texture. Being a grass rather than a grain, wild rice has fewer calories but more nutrients than ordinary rice. As wild rice can be expensive, it can be mixed with brown rice to make an attractive – although slightly more acid-forming – dish.

Winter Squash

Unlike summer squash, which have delicate skins and a short storage life, winter squash can be harvested very late into the autumn, have a longer storage potential and still provide an outstanding variety of nutrients. Winter squash is an excellent source of vitamin A through its betacarotene content, a very good source of vitamins C, B_1 and B_5, folic acid, potassium and dietary fibre. It is also a good source of vitamins B_6 and B_3.

Betacarotene has been shown to have powerful antioxidant and anti-inflammatory properties. A good intake of betacarotene can help reduce the risk of colon cancer, possibly by protecting colon cells from the damaging effects of cancer-causing chemicals. The anti-inflammatory effects may help reduce the severity of conditions like asthma, osteoarthritis and rheumatoid arthritis.

Varieties easily obtained are butternut and acorn squash. These make wonderful soups to give your meal a good alkaline start.

At-a-glance Look at Acid- and Alkaline-forming Foods

	FOODS TO AVOID	SECOND CHOICE of acid-forming foods – eat one small portion a day maximum	FIRST CHOICE of the best acid-forming foods – eat sparingly	THIRD CHOICE of foods – unlimited quantities	SECOND CHOICE of foods – unlimited quantities	FIRST CHOICE of foods – unlimited quantities
	ACID-FORMING FOODS			ALKALINE-FORMING FOODS		
FOOD CATEGORY	MOST ACID	ACID	LOWEST ACID	LOWEST ALKALINE	ALKALINE	MOST ALKALINE
VEGETABLES, BEANS & LEGUMES		Potatoes (without skins) Pinto beans Navy beans Lima beans	Cooked spinach Kidney beans String beans	Asparagus Onions Vegetable juices Parsley Raw spinach Broccoli Garlic	Okra Squash Green beans Beetroot Celery Lettuce Courgettes (zucchini) Sweet potatoes Carob	Carrots Tomatoes Fresh corn Mushrooms Cabbage Peas Potato skins Olives Soya beans Tofu
FRUITS	Blueberries Cranberries Prunes	Sour cherries Rhubarb	Plums Processed fruit juices	Lemons & limes Watercress Grapefruit Mangos	Dates, Figs Cantaloupe melons Grapes Papaya Kiwi, Berries Apples Pears Raisins	Oranges Bananas Cherries Pineapples Peaches Avocados
NUTS & SEEDS	Salted peanuts	Pecans Cashews	Pumpkin seeds Sunflower seeds		Almonds	Chestnuts
OILS			Corn oil	Olive oil	Flaxseed oil	Canola oil
GRAINS & CEREALS	Wheat White flour Cakes & pastries	White rice Corn Buckwheat Oats & Rye	Sprouted wheat bread Spelt Brown rice			Amaranth Millet Wild rice Quinoa

	FOODS TO AVOID	SECOND CHOICE of acid-forming foods – eat one small portion a day maximum	FIRST CHOICE of the best acid-forming foods – eat sparingly	THIRD CHOICE of foods – unlimited quantities	SECOND CHOICE of foods – unlimited quantities	FIRST CHOICE of foods – unlimited quantities
	ACID-FORMING FOODS			ALKALINE-FORMING FOODS		
FOOD CATEGORY	MOST ACID	ACID	LOWEST ACID	LOWEST ALKALINE	ALKALINE	MOST ALKALINE
MEATS & FISH	Beef Pork Shellfish	Turkey Chicken Lamb	Venison Cold-water fish			
EGGS & DAIRY	Cheese Homogenated milk Cream	Raw milk	Eggs (organic) Butter Yoghurt Buttermilk Cottage cheese		*Breast milk*	Soya cheese Soya milk Goat's milk Goat's cheese Whey
BEVERAGES (see Level 1)	Wines, beers & spirits Coffee	Decaffeinated or weak coffee	Black tea	Herb teas Green tea	Rooibos tea Ginger tea	Water Lemon water
SWEETENERS	NutraSweet Aspartame Sweet 'n' Low Hermesetas Candarel	White sugar Brown sugar	Processed honey Molasses		Maple syrup	Raw honey Raw sugar

BREAST MILK

The food nature intended for babies – breast milk – is alkaline forming. Without doubt the best food for human babies, breast milk contains the exact requirements for their growth and development. When breastfeeding, watch what you eat and drink as it will be passed to your baby in your breast milk. It is best to avoid smoking, drinking alcohol or a diet high in acid-forming foods.

The Best of the Acid-forming Foods

Foods from the following list should comprise 20 per cent of your daily intake. Some foods will become more acid forming during cooking, depending on the method used. The general rule is that the more something is cooked, the more acid forming it becomes.

When it comes to meal planning, even the best alkaline meal can be spoiled by what you accompany it with. For example, drinking alcohol, or having cheese and biscuits and/or coffee could make the entire meal acid producing, rather than the alkaline meal you had planned.

However, some acid-forming foods are excellent, providing valuable vitamins, minerals and amino acids. These foods must provide 20 per cent of your daily diet as many are required for building and repairing body cells. The best of the acid-forming foods are listed below.

There is also a list of acid-forming foods to avoid at the end of the chapter (*page 100*).

Bread – Seed Bread or Sprouted Wheat Bread

White bread is a refined food and is not allowed on the ph Diet. The refining process destroys most of the nutrients, leaving a lifeless food, high in gluten, yeast, calories and not much else. However, as bread plays a major part of most people's diet, and we don't all have time to make our own, we recommend Burgen® bread. This is cholesterol free, high in fibre and contains low levels of fat, salt and sugar. It has a low Glycaemic Index rating, which means it releases energy slowly. It contains soya and linseed for general wellbeing and has less wheat than most other breads.

Bharti's Bread (*see recipe on page 148*) contains low-fat cheese and chillies! Its protein content makes it a good addition to a salad or soup lunch. Corn and rye bread are other healthy alternatives. If you prefer wholegrain bread, choose the best possible loaf you can buy, preferably made in an old-fashioned bakery. Better still, make your own. But remember, all bread is acid forming, so go easy on it.

Cherries

Cherries are diminutive relatives of peaches and plums. Like their cousins, they have sweet, meaty flesh that surrounds a large pip. Cherries are an excellent source of vitamin C, but can sometimes be waxed, giving a false shiny appearance – always wash them well before eating. Avoid cherries with wrinkled skin or white spots that indicate mould.

Delicious eaten fresh, cherries also make a good choice to add to sorbets and yoghurts as a Level 3 dessert. According to a 1950 study of 12 people with gout, eating half a pound (about 227 grams) of cherries or drinking an equivalent amount of cherry juice prevented attacks of gout. The active ingredient in cherry juice remains unknown. We do know, however, that cherries are only mildly acid forming and as such help to maintain balance in our bodies.

Chicken

As chicken is such a popular food, many of you will be delighted to see it here. Although it is more expensive, try to purchase chicken that has been organically raised or is 'free-range', since these methods are more humane and produce chickens that are tastier and better for your health. Organic chickens have been fed an organically grown diet and have been raised without the use of hormones or antibiotics. Chicken is an excellent source of low-fat protein, vitamins B_3, B_5 and B_6, and the minerals selenium and phosphorus.

Cottage Cheese

Cottage cheese is a soft, fresh-curd variety of cheese that has been made in Europe and America for centuries. It is an uncured cheese (one that has not been aged), and is favoured by weight-watchers because it is lower in fat than most cheeses. Cottage cheese is a good source of low-fat protein, as well as being an excellent source of selenium and vitamin B_2 (riboflavin), and a good source of calcium.

An easy food to eat, cottage cheese is always good to have in the refrigerator for emergencies. It is an excellent lunch choice and goes well with fruits, such as pineapple, peaches or berries. Add your own fruit rather than choosing cottage cheese with fruit already added. As cottage cheese is slightly acid forming, having it with tomatoes and avocados will balance it up nicely. In some recipes, it can also be substituted for a fuller-fat cheese.

Eggs

Eggs are an excellent source of low-cost but high-quality protein. People tend to avoid eggs because of bad publicity regarding cholesterol and salmonella. However, many studies have found that it is saturated fat in the diet, not dietary cholesterol, that influences blood cholesterol levels the most. To destroy any bacteria there may be, eggs should be cooked at high enough temperatures for a sufficient length of time. Soft-cooked, sunny-side-up or raw eggs carry a risk. Hard-boiled, scrambled, poached or baked eggs do not.

Eggs are an excellent source of vitamin K, and a very good source of vitamin B_{12}, iodine, vitamin D, selenium plus many other valuable nutrients. The egg is such an excellent source of complete protein that it is used as a standard for all other proteins to be measured against. Try to purchase eggs that come from chickens fed flaxseeds in their diet, as they will be an excellent source of omega-3 fatty acids. These will include most organic and free-range eggs, but check labels carefully. If you only buy one organic product, make sure it is eggs.

Fish (cold-water fatty fish)

Salmon, trout, mackerel, herring, sardines and other cold-water fatty fish are low in calories and saturated fat yet high in protein and omega-3 essential fatty acids. As their name implies, essential fatty acids are *essential* for human health. However, because the body cannot make them, they must be obtained from foods in the diet on a daily basis. Wild-caught cold-water fish are higher in omega-3 fatty acids than warm-water fish. In addition to their high concentration of omega-3s, they are an excellent source of the vitamins B_{12} and B_3, and the trace mineral selenium.

Cold-water fatty fish have long been considered 'brain food'. This is because of their high concentration of omega-3 fats. The human brain is made up of more than 60 per cent fat. For brain cells to function properly, this fat needs to be primarily omega-3 fats, such as the EPA (eicosapentaenoic acid) found in salmon and all oily fish.

Kidney Beans

Like most beans, kidney beans are an excellent source of choles-terol-lowering fibre. The high fibre content also prevents blood sugar levels from rising too rapidly after a meal, making these beans an especially good choice for individuals with diabetes, insulin resistance or hypoglycaemia. When combined with whole grains such as rice, kidney beans provide a virtually fat-free, low-calorie but high-quality complete protein.

Kidney beans are an excellent source of the trace mineral molybdenum, which is required to make the enzyme responsible for the detoxification of sulphites in the body. Sulphites are a type of preservative commonly added to prepared foods like delicatessen and salad-bar salads. Kidney beans are also an excellent source of folic acid, vitamin B_6, manganese, magnesium and zinc. In addition to providing slow-burning complex carbohydrates, kidney beans can increase your energy by helping to replenish your iron stores. This makes them particularly valuable for menstruating women.

Lamb

Although a source of saturated fat and cholesterol, red meat can benefit our diet. Organic English lamb is probably one of the best meats to choose. Lamb is an excellent source of selenium, zinc, some B-vitamins including B_{12}, protein and iron. Vitamin B_{12} supports production of red blood cells and prevents anaemia, allows nerve cells to develop properly and helps your cells metabolize protein, carbohydrate and fat.

The trace mineral zinc affects many fundamental processes, perhaps the most important of which is immune function. If one mineral were singled out for its beneficial effects on the immune system, it would have to be zinc. It is critical not only for immune function but also for wound healing and normal cell division. Zinc also helps stabilize blood sugar levels and the body's metabolic rate, and is necessary for our sense of smell and taste to work at their best.

To benefit from these nutrients, a 115-gram (4-ounce) portion of lamb is all that is required. As with all acid-forming proteins, quality is more important than quantity.

Lentils

Lentils are particularly popular in India where they are cooked to a purée and called dhal. They make great soups and can be used in casseroles, salads and stews; they also make excellent croquettes or patties. They are fast and simple to prepare, and make a nourishing, hearty and inexpensive meal when eaten with brown rice – making a complete protein. Although it is not necessary to soak lentils, they should be thoroughly picked over and washed to remove impurities. Lentils are an excellent source of iron, potassium and folic acid and a good source of vitamin B_3. They make an excellent, healthy substitute for refined white pasta.

Lima Beans

Sometimes called 'butter beans' because of their starchy yet buttery texture, lima beans have a delicate flavour that complements a wide variety of dishes. Like kidney beans and most other legumes, lima beans are an excellent source of cholesterol-lowering fibre. They are also an excellent source of molybdenum (see 'Kidney Beans', page 94).

The lima bean's contribution to heart health lies not just in its fibre, but also in the significant amounts of folic acid and magnesium it supplies. Folic acid helps lower levels of the toxin homocysteine. When enough magnesium is around, veins and arteries breathe a sigh of relief and relax. This lessens resistance and improves the flow of blood, oxygen and nutrients throughout the body. In addition to all this, lima beans can increase your energy levels by helping to replenish your iron stores. A cup of lima beans contains nearly 30 per cent of the daily requirement for this important mineral.

Lima beans are best bought in their dried form and soaked overnight before using. They make great starters, main meals and hearty soups, and are an important staple in the pH Diet.

Nuts

Nuts are one of the food groups we tend to avoid because of their high fat and calorie content. However, they are packed with nutrients, which is more important than their calorie or fat content. Nuts reduce levels of harmful LDL cholesterol, help maintain strong teeth and bones, lower the risk of coronary artery disease, boost levels of omega-3 fatty acids, reduce the effects of free radicals, increase brain power and concentration and support the immune system. They make an easy, highly nutritious snack food, but they can also be ground into milk to relieve intestinal spasm and inflammation in cases of irritable bowel syndrome. They are ideal for sprinkling over breakfast cereal to make a more balanced meal.

Almonds (see alkaline list)

Brazil Nuts
Brazil nuts contain up to 250 times more selenium than most foods, making them high in antioxidants. They also contain vitamin E, which helps support the immune system. Essential fatty acids assist in lowering harmful LDL cholesterol levels and reducing the effects of free-radical damage.

Chestnuts

Chestnuts contain vitamins B and C and the minerals iron, magnesium, potassium and zinc. Particularly beneficial for the blood and circulatory system, chestnuts also relieve indigestion. Chestnuts are a good nut to choose for the elderly and for convalescents, as when cooked they become soft and easy to eat. They also benefit those prone to varicose veins and piles.

Hazelnuts

Hazelnuts are high in calcium, phosphorus, copper, iron, magnesium, potassium, sulphur and essential fatty acids. They also contain good levels of folic acid, making them a healthy snack in pregnancy. Their high boron levels not only assist the absorption of calcium but also make them an exceptional brain food, stimulating electrical activity in the brain. Essential fatty acids and fibre make hazelnuts good as a snack or for adding to breakfast cereals for balance and extra energy. Hazelnuts are recommended for people who are prone to stones in the kidney or gall bladder. They may also help to dispose of intestinal worms – take 1 tablespoon of cold-pressed hazelnut oil every morning on an empty stomach for 15 days.

Peanuts

Salted or dry-roasted peanuts do not have the same beneficial effects as fresh peanuts taken from the shell. Fresh peanuts contain vitamins B and E, trace elements, amylase and the antioxidant resveratrol. Peanuts are beneficial for the digestive system, particularly for those suffering from indigestion. The enzyme amylase eases this condition.

Peanuts are believed to be helpful at lowering your risk of suffering a heart attack or stroke. The powerful antioxidant resveratrol, also found in red wine, prevents the oxidation of harmful LDL cholesterol, a process that results in arteries furring up. Peanuts are also effective regulators of insulin and blood sugar, making them a good choice for anyone suffering from glucose intolerance or diabetes.

Walnuts
Walnuts contain vitamins A, B and C and the minerals copper, iron, magnesium, potassium, selenium and zinc. They have the highest amounts of omega-3 fatty acids of all nuts. This helps prevent conditions such as psoriasis, eczema, arthritis and mental illness. Walnuts are also reported to be good for the elimination of intestinal parasites and for alleviating heart and circulatory problems.

Oats

A great way to start the day, oats are an excellent source of a special type of soluble fibre known as beta-glucan. This has been shown to lower cholesterol levels and have beneficial effects in diabetes. Oats raise blood sugar more slowly than white rice or bread, so starting the day with a bowl of oats may make it easier to keep blood sugar levels under control for the rest of the day. By adding nuts, seeds and fruit to the oats, you have a balanced breakfast that will keep you going for hours. Oats are an excellent source of manganese, and a very good source of selenium, phosphorus and tryptophan.

Plums

Relatives of the peach, nectarine and almond, plums belong to the *Prunus* genus of plants. They are 'drupes' fruits with a hard stone surrounding their seeds. When plums are dried, they are known as prunes. Plums have been the subject of research for their high content of unique phytonutrients called phenols, and their function as antioxidants has been well documented. The vitamin C in plums provides antioxidant protection and is thought to help the body absorb iron. Plums are also a good source of vitamins B_1, B_2 and B_6, vitamin A in the form of betacarotene, vitamin E and dietary fibre. Plums or prunes are an excellent breakfast choice for the pH Diet. They make delightful desserts and can also be added to soups and used to make bread.

Pumpkin Seeds

Pumpkin seeds are nutritious and tasty. They are a very good source of phosphorus, manganese, magnesium, copper, tryptophan, zinc, iron and protein. Like any seeds, pumpkin seeds can be eaten at any time for a quick snack, but there are many other ways of utilizing these amazing, health-promoting little seeds. You can add pumpkin seeds to sautéed vegetables to give a bit of 'crunch'; you can sprinkle them on top of mixed green salads; and you can grind them up, together with other seeds, to sprinkle over breakfast cereals or into soups. Be sure to make the tasty salad dressing in the recipes section, which contains ground seeds, fresh garlic, parsley and coriander leaves – delicious (*see page 123*).

Rice (brown)

The difference between brown and white rice is not just the colour. A whole grain of rice has several layers. Only the outermost layer, the hull, is removed to produce what we call brown rice. This process is the least damaging to the nutritional value of the rice and avoids the unnecessary loss of nutrients that occurs with further processing. If brown rice is further milled to remove the bran and most of the germ layer, the result is whiter rice, but also rice that has lost many more nutrients. Polishing the rice removes even more health-supporting nutrients. The complete milling and polishing process that converts brown rice into white rice destroys 67 per cent of the vitamin B_3, 80 per cent of the vitamin B_1, 90 per cent of the vitamin B_6, half the manganese, half the phosphorus, 60 per cent of the iron, and all of the dietary fibre and essential fatty acids!

Brown rice is a very good source of manganese, a good source of the minerals selenium, magnesium and phosphorus as well as the vitamins B_6, B_3 and B_1 and dietary fibre. Only brown rice or basmati rice, which is a naturally occurring white rice, are recommended on the pH Diet.

Rye

For the many people who are sensitive to wheat, rye is a useful alternative. Rye is a very good source of fibre, the amino acid tryptophan, copper and phosphorus. One of the most important properties of fibre is its ability to bind to toxins in the colon and then remove them from the body.

Rye may also be helpful for women going through the menopause. It contains a substance that can affect the amount of oestrogen produced by the body. For some women, the phytoestrogens in rye are just strong enough to help prevent or reduce uncomfortable symptoms that may accompany menopause, such as hot flushes, which are thought to be due to plummeting oestrogen levels. On the other hand, when too much oestrogen is around, rye can help lower oestrogen and provide protection against breast cancer.

Sunflower Seeds

Sunflower seeds are an excellent source of vitamin E and a very good source of vitamins B_1 and B_5, copper, manganese, magnesium, selenium and phosphorus. They are also good sources of tryptophan, zinc and vitamin B_6.

Vitamin E travels through the body neutralizing free radicals that would otherwise damage fat-containing structures and molecules, such as cell membranes, brain cells and cholesterol. By protecting these components, vitamin E has significant anti-inflammatory effects that can reduce the symptoms of asthma, osteoarthritis and rheumatoid arthritis. Vitamin E has also been shown to reduce the risk of colon cancer, help decrease the severity of hot flushes in women going through the menopause and help reduce the development of diabetic complications.

Being a good source of magnesium, sunflower seeds help reduce the severity of asthma, lower high blood pressure and prevent migraine headaches, as well as reducing the risk of heart attack and stroke.

Tortillas

Tortillas are the traditional bread of Northern Mexico and have been included in the pH Diet for their versatility. Either make your own or buy good quality tortillas, checking food labels for a product containing the least number of E numbers and salt that you can find. They are perfect for a satisfying breakfast, lunch or snack, stuffed with a selection of alkaline-forming salad foods. Check the recipe section for a few ideas. Drizzle savory wraps with Essential Balance and a tablespoon of '3&6 Mix' for balance, before wrapping up the tortilla. You can also wrap up fruit and yoghurt for a breakfast, desert, or snack – adding '3&6 Mix' for crunch and balance. Traditionally tortillas are made with wheat flour, so avoid if you are wheat sensitive.

Turkey

Once kept for special occasions like Christmas, turkey is now freely available all year round. It is an excellent source of low-fat protein and of selenium, zinc and important B-vitamins. Selenium is of fundamental importance to human health. It is an essential component of thyroid hormone metabolism, antioxidant defence systems and immune function.

Turkey is also a good source of the vitamins B_3, B_6 and B_{12}. These three B-vitamins are important for energy production. Vitamin B_6 is essential for the body's processing of carbohydrate, especially the breakdown of glycogen, the form in which sugar is stored in muscles.

The same rule applies for turkey as with other meats – go for quality rather than quantity. You only need a small portion of turkey to balance a meal.

Yoghurt

Yoghurt is an ancient wonder food, strongly anti-bacterial and anti-carcinogenic. Eating 225 grams (8 ounces) every day can boost the immune system considerably. As well as the beneficial culture yoghurt contains – the probiotics – it is also a good source of absorbable calcium. Yoghurt maintains a healthy digestive tract; combats yeast overgrowth and fungal infections such as *Candida albicans*; prevents constipation, diarrhoea, flatulence and bloating; protects against osteoporosis and rheumatoid arthritis; overcomes skin problems; lowers cholesterol levels; and enhances general nutritional status.

Probiotics means 'for life', the opposite of antibiotics, which means 'against life'. Even one course of antibiotics destroys the beneficial as well as the harmful bacteria in our intestines, resulting in a suppressed immune system. Many babies and children are given multiple courses of antibiotics, some before they are even a year old. Although antibiotics can be life-saving, you should always follow them up with a course of probiotics. These promote the growth of 'friendly' bacteria in our intestines, and thousands of studies have confirmed that probiotics are indispensable in keeping us healthy.

The good flora in our small intestines make B-vitamins, such as biotin, B_3, B_6 and folic acid. They provide the enzyme lactase, which helps us digest dairy foods, including the vital calcium they contain. They also help considerably to enhance bowel function. When friendly bacteria are absent from the bowel, it takes longer for food to pass completely through the system.

Unfortunately, it is impossible to know exactly how many live cultures are in the yoghurt you are eating. A pot may only provide a couple of hundred bacteria. This is fine if you are just keeping 'topped up' with friendly bacteria, but if you have been on a course of antibiotics, and especially if you have had several courses, you need to replace the billions of destroyed bacteria as soon as possible. These need to come in the form of a good probiotics supplement. Once opened, the probiotics should be refrigerated. Children's formulas are available from all good health-food stores.

The Best Way to Eat

You have probably noticed that the majority of the foods listed in this chapter are vegetables and fruits. It is easy to juice them or use them in soups. However, to retain their beneficial alkalizing effects, these foods should be cooked as little as possible. Although juices are of great nutritional value, it is important to eat the whole fruit and vegetable where possible, as it is the fibre contained in the skins that binds to the toxic waste and removes it from the body.

Now we have looked at the top 80 alkaline-forming foods, and the best 20 acid-forming foods, how are you going to introduce them into your daily diet?

Variety is essential. We tend to eat the same meals day after day and rarely try anything different. Most of the menus listed in the recipes section (Chapter 3) are quick and easy to put together, although one or two may present a bit of a challenge. Go for it – variety is the spice of life.

Acid-forming Foods to Avoid

- Meats: beef, pork, bacon, sausages, burgers, pies, ham, faggots, all processed meats
- All refined, packaged foods including most breakfast cereals that contain sugar
- All ready-made meals, including 'diet' foods, 'reduced-fat' foods and '99% fat-free' foods
- Wines, beers and spirits
- Carbonated drinks
- Cow's milk
- Fried food
- Green bananas
- Cranberries
- White bread
- White flour and sugar
- Battery eggs
- Ice cream
- Margarine
- Salt
- Biscuits and cakes
- Snack food
- Refined pasta
- Barbecue food

CHAPTER THREE

Recipes

With time being precious for most of us due to our hectic lifestyles, the following recipes are quick to prepare. Some will be easy to make while others may be a little more challenging, but all are a joy to eat! All of them will increase and maintain the vital alkaline reserves in your body.

Accompaniments like Bharti's Bread (*page 148*) and Bharti's Special Dhal (*page 124*) give depth and sustenance to soup and salad dishes, and are great fun to make. Delicious recipes for fruit and vegetable juices are well worth the effort and are your ticket to looking good and being well. Think of getting back into the kitchen as your new, enjoyable hobby.

Notes on the Ingredients

As the pH Diet is specifically designed to boost the alkaline-producing foods in your body, there are a few general notes about ingredients and methods that you will find useful to read before you start to cook.

Tomatoes

Raw tomatoes are alkaline forming because of their low sugar and high water content. When cooked, however, they become moderately acid forming. Cooked tomatoes will therefore be more evident in Level 3 of the programme when alkaline reserves have been built up. Raw tomatoes are used more in Level 2 when you are building up reserves.

Alkaline-forming Super Foods

Recipes including complex carbohydrates (pulses, lentils, peas and beans), which may become more acid forming through cooking, and proteins (meat, chicken, fish, eggs or cheese), which are acid-forming foods, can be improved immensely by including any super alkaline-forming food from the 1st choice list (*see Chapter 2, pages 88–9*). Foods such as raw tomatoes, avocados, and lime or lemon juice will help redress the balance.

Oils

Use only cold-pressed oils such as flaxseed, olive or borage oil, or a combination like Udo's Choice or Essential Balance (*see Resources*) to protect your body from acid waste. Do not heat these oils.

Healers and Cleansers

Remember that vegetables are used for healing and building the body in Level 2, and that many fruits are slightly acid forming and are used as cleansers in Level 3. We heal before we cleanse. By healing the body first, it is better able to cope with detoxifying the waste.

Eating Out

It is useful to learn and memorize the recommended foods, as this can be a great help when eating away from home. Water (neutral), limes, tomatoes and avocados are alkaline super foods, and always a good choice. Water will dilute acid wastes and assist in carrying them out of the body, while tomatoes and avocados will help balance the acid in the rest of the meal, and provide the minerals needed to neutralize the proteins.

Dairy Products

Milk is not only acid forming but also mucus forming, and has long been associated with chronic catarrh, sinusitis and hay fever. People in the UK and the US drink more milk than probably any other country, yet we have the highest rates of osteoporosis. As milk is very acid forming, the body removes calcium from the bones to maintain alkalinity. So if you eat and drink excess dairy produce on a daily basis, then you are probably losing calcium from your bones on a daily basis too! The long-term result of this may well be osteoporosis.

As mentioned earlier, vitamin D and magnesium as well as calcium are needed for strong bones. As magnesium is found in all alkaline-forming green leafy vegetables, they will keep calcium in the bones, where it belongs.

Take care not to eliminate milk and other dairy products from your diet without using an alternative. We recommend a gentle change over to more soya milk and soya products.

Nut and Seed Milks

As you decrease your intake of dairy produce and become accustomed to drinking soya milk and other milks, you may want to try your hand at making your own nut and seed milks. These are wonderful for adding a creamy texture to your dressings and soups, or just to drink for a snack. Their richness and sweetness can be varied, depending on how you dilute them. They are also good sources of protein and calcium. You can now buy these nut milks in cartons from health-food shops, but as with most things, nothing beats making your own.

Eggs

Eggs are in the list of best acid-forming foods (*page 88–9*), but this applies only to *organic* eggs. Please remember that acid-forming foods are not forbidden on the pH Diet. We are looking for a balance of acid- and alkaline-forming foods, so eggs are allowed. Organic eggs, particularly Columbus eggs, are an excellent choice, as the chickens are given high quantities of flaxseed in their diet, the benefits of which are passed directly to you.

The lecithin in eggs breaks down cholesterol in the body, provided the eggs are boiled, poached or baked. Eggs only become a high-cholesterol food when they are fried, and the lecithin is destroyed. So, as with all good foods, choose quality over quantity and enjoy the occasional boiled egg. If you only buy one organic food, make it eggs.

Fruit

It is best to eat fruit alone or at the beginning of a meal as it is so readily digested. Whatever goes into the stomach first is digested first, and sets the pace for the rest of the meal. Fats and proteins are digested much more slowly than most fruits and vegetables. If fruits are eaten after bread, butter and beans, for example, they are forced to wait in the stomach for as long as it takes for these heavy foods to be digested. This delay in the hot, wet, acid environment of the stomach causes fermentation, gas and belching.

Sprouting

Sprouted seeds are rich in vitamins, and a clean, cheap, uncontaminated food source. Bean sprouts, particularly mung beans, have been grown in the Far East for thousands of years. Common seeds for sprouting are alfalfa, mung beans, aduki beans, wheat, barley, fenugreek, lentils, mustard, oats, pumpkin seeds, sesame seeds, sunflower seeds and soya beans. Most supermarkets now sell ready-sprouted seeds.

Incredibly versatile, sprouted seeds are best eaten raw in salads, sandwiches or on their own. They can also be added to hot dishes such as soups or casseroles. This should be done at the last minute to maintain their freshness and nutritional value.

Sprouting your own seeds is a quick, easy and incred-ibly cheap way of providing organic, fresh food bursting with vitality. A mere tablespoon of alfalfa seeds will produce about 1 kilogram (over 2 pounds) of sprouts. Sprouted seeds are an almost perfect food, the richest-known source of naturally occurring vitamins. Additionally, the sprouting process 'pre-digests' the seed, making it much easier for us to break down and absorb its goodness. This is the reason why many sprouted grains and legumes are less likely to cause the allergic reactions their non-sprouted counterparts can trigger.

Sprouted mung beans contain as much as 120 milligrams of vitamin C per 100 grams compared to only around 53 milligrams in oranges. The amount of B-vitamins rises as a seed sprouts. Sprouts may contain vitamin B_{12}, and the fat-soluble vitamins A, D and E are also present. Minerals are in abundant supply in sprouts. Unlike most vegetables, sprouts are also a good source of protein. Soya sprouts are the only ones to contain all eight essential amino acids.

How to Sprout

Small commercial sprouters are available in health-food stores. They usually comprise four small trays stacked upon each other with a selection of seeds and beans and a set of instructions. Alternatively, you can use a standard seed tray, a colander or a sieve.

1. Place a handful of your selected seeds or beans in a bowl, cover with cold water and leave to soak overnight. Most seeds will expand up to eight times their size so be generous with the water, submerging them completely.
2. The following morning, drain off the water, rinse well and lay the seeds in the base of the sprouter.
3. Either leave on a windowsill or place in an airing cupboard. The seeds need to be rinsed thoroughly at least once, and preferably twice, a day.
4. Continue to rinse the seeds daily until ready to harvest. After about three to five days you should have a crop of sprouts. Once they are the right size for eating, give them one final rinse and store them in the fridge to stop them from growing any further.

Sprout harvesting times

SEED/SPROUT	HARVEST TIME IN DAYS
Aduki beans	4–6
Alfalfa	5–7
Barley	3–4
Chickpeas	4
Fenugreek	4–5
Flageolet beans	3–5
Green lentils	3–5
Green peas	3–5
Mung beans	2–3
Radish	4–5

Rye	3–5
Soya beans	3–6
Sunflower seeds	4–6
Wheat	2–4

Breakfasts

Breakfast is the most important meal of the day and an integral part of the pH Diet. This is the meal that will probably mean the most changes for you as it involves reducing tea, coffee, sweet cereals, cow's milk and wheat products. You will, however, enjoy alternative grains, soya milk, fruits for cleansing and the '3&6 Mix' for essential fatty acids. You can make this yourself or buy it from health-food shops (*see Resources*). All the following breakfast recipes are alkalizing, and will give you an excellent start to the day.

Almond Milk

Makes approximately 250ml (9fl oz/1¹/₄ cups)

This milk will keep for three to four days. It is great on hot grains such as quinoa, buckwheat, millet or amaranth. You can thin it further with more water if you want to.

    ¹/₂ cup almonds (soaked for 12 hours)
    ¹/₂ cup pine nuts (soaked for 6 hours)
    1 cup spring or filtered water

1. Put the soaked almonds and pine nuts in a blender and pulverize.
2. Add the water gradually, while continuing to blend on high.
3. Strain through a fine sieve or cheesecloth.

'3&6 Mix'

Makes 60 level tbsp

490g (18oz) golden linseeds
170g (6oz) pumpkin seeds
170g (6oz) sesame seeds
170g (6oz) sunflower seeds

Use organic seeds whenever possible. There is some debate as to whether the seeds are the most beneficial left whole or ground immediately before use. Grind half your daily allowance and leave the other half whole. Alternatively, grind the seeds one day and leave them whole the next. Always grind immediately before use as, once ground, the seeds are exposed to air and can oxidize.

1. Mix all the ingredients together and put into a sealed jar. Shake well.
2. Before serving, grind 1 tbsp of seeds in a coffee grinder for a second or two.
3. Sprinkle into soups or over vegetables, and add to sandwiches or to breakfast cereals and other recipes.

Breakfast Rice

Serves 1

You can use leftover rice in this recipe.

1 cup basmati rice, cooked
1 cup milk (goat's or soya)
$1/4$ cup raisins
$1/4$ tsp cinnamon

1. Put all the ingredients into a small saucepan. Mix well.
2. Cook on a medium heat for about 5 minutes until hot.

Stewed Figs or Prunes

Serves 1

This excellent cleanser makes a delicious breakfast dish or a good starter.

    6 dried figs or prunes
    1 cup boiling water

1. Place the figs or prunes in a heat-proof cup or bowl.
2. Pour boiling water over the dried fruit.
3. Let it soak for 15 minutes, covered if possible.
4. Drain and serve with a small pot of bio yoghurt.

Fresh Fruit Salad

Serves 3–4

This makes a good breakfast dish or a starter for summer lunch or dinner.

    1/2 cup fresh sweet blueberries
    1/2 cup sweet apricots
    1/2 cup peaches
    1/2 cup fresh strawberries, halved
    1 tbsp lemon or lime juice

1 tbsp honey (optional)

Mint leaves for garnish (optional)

Small pot of bio yoghurt (optional)

1. Chop larger fruit into $1/2$ –1-inch pieces.
2. Put all fruit into a serving bowl.
3. Drizzle juice and honey over the fruit.
4. Mix well and serve, garnished with mint leaves and drizzled with yoghurt, as desired.

Millet Porridge

Serves 1

100g (3^1/$_2$oz) millet flakes

140ml (^{1}/$_4$ pint/ ^{2}/$_3$ cup) milk (rice, goat's or soya) or water

1 dessertspoon '3&6 Mix'

1. Place the millet in a large, microwave-proof cereal bowl and cover with either milk or water.
2. Microwave on full power for 1 minute, remove and stir well. Cook for another minute in the microwave and remove.
3. Sprinkle over '3&6 Mix' and serve with additional milk if required.

Weekend Breakfast

Serves 1

You may have a little more time to prepare a cooked breakfast at weekends. A boiled egg is a really good way to start any day. You can eat up to 6 eggs per week. Serve your weekend breakfast with a glass of fresh orange juice.

1 Columbus egg, boiled or poached
1 slice of rye or Burgen bread with a little organic butter or ghee

Papaya Salad

Serves 1

This delicious, colourful salad is very good for stimulating digestion before a protein meal and for clearing excess mucus out of the digestive tract. It is best eaten on an empty stomach.

1 ripe papaya
Sea salt to taste
1 tbsp lime juice

1. Wash the papaya and slice it in half, removing the seeds. Cut into wedges, removing the peel as you go. Cut the wedges into 2.5cm (1-inch) pieces, or whatever size you like.
2. Salt the papaya lightly but thoroughly and pour the lime juice over it all.

Bharti's Ginger Mango

Serves 1

I can eat mango any time. When you are feeling low, try slicing green mango and sprinkling it with salt, sugar and a little bit of chilli. Wow! For a real treat, roll a ripe mango in your hands until it is very soft and you can feel the stone move. Then cut a hole in the top and very slowly enjoy the fruit that you can press out. It's a joy. You can also try this easy recipe.

1 ripe mango
A little organic butter or ghee
Ground ginger to taste
Salt to taste

1. Peel and slice a ripe mango.
2. Place the slices in a sauté pan containing a little melted organic butter or ghee. Warm the slices on one side, then turn and warm the other. The flesh should be just beginning to brown at the edges.
3. Remove the fruit to a dessert plate and sprinkle with ground ginger and salt.
4. This can be served as a starter, during the main meal or as a dessert.

Poached Peaches

2 ripe peaches
1 vanilla pod
1.2 litres (2 pints/5 cups) water
225g (8oz) frozen raspberries or other red fruit
55g (2oz) slivered almonds (to serve)

1. Parboil the peaches in boiling water for 1 minute. Peel and leave whole.
2. In a saucepan, boil the vanilla pod in the water for 3–4 minutes. Remove the pod and poach the fruit over a low heat for 5 minutes. Transfer to a serving dish.

3. Purée the raspberries in a blender. Pour them over the peaches and serve with slivered almonds.

Soups

Soups are an important part of the pH Diet. They make excellent, nourishing meals that are usually easy to prepare. You can make them ahead of time and warm them up for a hearty lunch, dinner or even breakfast, if you so desire. Clear soups are easy to digest. Cooling in spring and summer and warming in autumn and winter, they also serve as a support in convalescence and older age. They offer a good way to begin to heal the digestive and nervous systems, which may have been traumatized or neglected with too much fast food eaten on the run.

All soups should be homemade if time allows. Level 2 soups are cooked through and partly blended. If you are short of time, shop-bought fresh soup is allowed, as long as it is free of additives.

Level 3 soups are cooked far less in order to keep their enzymes and vitamins 'alive'. The more raw the soup, the more alkaline forming it will be.

Melon and Peach Summer Soup

Serves 2

1 large cantaloupe melon
4 fresh, soft peaches, peeled
Lemon juice (optional)

1. Purée the melon flesh and peeled soft peaches in a blender or food processor to make delicious cold soup.
2. Add lemon juice to taste if required.

Quick Broccoli Soup

Serves 2

1 medium head broccoli
1 medium onion, quartered
230ml (8fl oz/1 cup) water
1 vegetable stock cube
225g (8oz/1 cup) Greek-style yoghurt
1 tsp ground cumin
Freshly ground black pepper to taste
Fresh coriander (cilantro), finely chopped

1. Remove the outer leaves from the broccoli and cut off most of the stem. Wash the remaining head and break into pieces.
2. Combine the broccoli, onion, water and stock cube in a large saucepan. Bring to the boil, cover and cook over a medium heat until the broccoli is tender – approximately 10 minutes.
3. Carefully transfer the vegetables and stock to a blender. Add the yoghurt and cumin. Cover and blend. Adjust the consistency of the soup with a little soya milk if it is too thick for your liking. Season with black pepper, if necessary.
4. Return the soup to the pan if you plan to reheat it before serving, or simply pour into soup plates and sprinkle with the fresh coriander.

Broccoli Sunflower Soup

Serves 4

This soup makes a good quick lunch on a cold day.

 4 cups broccoli, chopped (about 1 large head)
 2 cups water
 1/4 cup raw sunflower seeds, shelled
 1 tbsp miso
 2–3 cloves of garlic
 2 spring onions (scallions), finely chopped
 1/2 tsp dried oregano
 Black pepper to taste

1. Put the broccoli in a steamer over the water. Cover and steam until tender and bright green – about 5 minutes.
2. Meanwhile, grind the sunflower seeds into a fine powder in the blender and leave them there.
3. When the broccoli is done, put it in the blender with the cooking water plus the rest of the ingredients and purée with the ground sunflower seeds. Serve immediately.

Pea and Sweetcorn Soup

Serves 2

Serve piping hot as a starter or with Bharti's Bread (*page 148*) for a complete meal.

 1 tbsp corn oil
 1 tsp cumin seeds

1 tsp fenugreek

1 large clove of garlic, minced

1 green chilli pepper, seeded and finely chopped

1 x 425g (15oz) can plum tomatoes

1/2 cup frozen garden peas

1 cup frozen tender sweetcorn

1 vegetable stock cube made up to 1/2 litre (1 pint/2 1/2 cups) with boiling water

1/4 small head cabbage (green or white), finely chopped

Black pepper to taste

1. Combine the oil, cumin seeds and fenugreek in a large saucepan and heat until the seeds sizzle.
2. Lower the heat, add the garlic and chilli and cook until the garlic begins to brown.
3. Add the tomatoes, peas, corn and stock. Stir.
4. Place the cabbage on top of the combined ingredients, cover the pan and simmer for 20 minutes. Add black pepper to taste.

Starters

Starters can be the best part of the meal. How many times have you been in a restaurant and fancied many of the starters and none of the main courses? Many restaurants will allow you to select two starters instead of one and a main course. This works well because you will probably consume far fewer calories. This concept can work at home too. Rather than make one big meal, make two smaller ones instead, choosing delicious ingredients for variety. Have a rest between the two dishes to give your stomach time to receive messages from your brain that you are full.

Avocado and Tomato Cleanser

Serves 1

This attractive starter is an excellent way to start any meal.

  1 large, ripe avocado
  1 medium tomato, sliced
  Udo's Oil or Essential Balance

1. Peel the avocado, remove the stone and slice into segments.
2. Arrange slices of avocado and tomato in a decorative way.
3. Drizzle with Udo's Oil or Essential Balance or add a couple of olives.

Lima Beans and Onions

Serves 2

This makes a really quick starter or, if you increase the ingredients, a satisfying lunch.

  1 x 425g (15oz) can lima beans
  1 large onion, sliced into rings
  Fresh parsley
  Udo's Choice, Essential Balance or olive oil to dress

1. Open the can of beans and drain well.
2. Mix the onion rings with the beans and arrange on an attractive plate.
3. Roughly chop the parsley and sprinkle over the beans and onions.
4. Drizzle with oil to taste.

Prawn and Pineapple Cocktail

Serves 4

1 ripe pineapple, diced
115g (4oz) large cooked prawns
Fresh ginger, grated
Olive oil
Sea salt and freshly ground black pepper to taste
8 leaves of romaine lettuce

1. Combine the pineapple with the prawns, ginger and a little olive oil.
2. Season to taste and serve on a bed of romaine lettuce.

Guacamole

Serves 2

This really alkaline-forming recipe is a good start to any meal.

1 large, ripe avocado
1 tomato, finely chopped
1/4 tsp sea salt
Juice of 1 lime
Chilli powder to taste

1. Mash the avocado and mix with the other ingredients.
2. Use as a starter with crudités – raw sweet (bell) peppers, celery, aubergine (eggplant), cucumber or summer squash.

Corn on the Cob

Serves 4

Do not add salt to the boiling water when cooking corn kernels as it makes them go hard, and take care not to overcook them.

4 corn kernels

1. Drop the corn kernels into deep, boiling water and cook for approximately 8–10 minutes, depending on size.
2. Eat the cobs either just as they are or seasoned with a little organic butter, olive or flaxseed oil, salt and pepper, or any herb or spice you enjoy.

Pumpkin Seed Salad Dressing

Serves 2

30g (1oz) pumpkin seeds, ground
1 clove of garlic, crushed
Good sprig of parsley, finely chopped
Good sprig of coriander (cilantro), finely chopped
1 lemon, squeezed
60ml (2fl oz) olive oil

1. Mix all the ingredients together in a screw-top jar.
2. Shake well – delicious!

Main Meals

The earlier in the evening you have your main meal, the better it is for your digestive system. In fact, a main meal at midday is better still – but unfortunately this is not possible for many people. Do your best not to eat under stressful conditions. Take your time eating – a starter followed by a rest then a small meal is much more beneficial than rushing a large meal.

Bharti's Special Dhal

Serves 4

This is a really easy dish to make – don't let the long list of ingredients put you off.

    1 x 425g (15oz) can cooked chickpeas (garbanzo beans)
    1 x 425g (15oz) can cooked soya beans (soy beans)
    Squeeze of lemon juice
    3 tbsp corn oil or ghee
    1 tsp cumin seeds
    1 tsp ground coriander (cilantro)
    1 large onion, finely chopped
    1 large red (bell) pepper, seeded and chopped
    1 tsp fresh root ginger, finely chopped
    2 large garlic cloves, minced
    1 large green chilli pepper, seeded and finely chopped
    Freshly ground black pepper

1. Empty both cans of beans into a large sieve and rinse well under cold running water. Tap dry and, with the beans still in the sieve,

sprinkle with the lemon juice, shaking gently to distribute it evenly.

2. Place the oil or ghee in a heavy pan with the cumin seeds and coriander and cook over a medium heat until the seeds pop and dance in the pan. Add the onion, pepper, ginger, garlic and chilli, and cook until the onions are translucent. Remove the pan from the heat.

3. In a separate pan, combine the beans with three or four tbsp of water and bring to the boil. Lower the heat to a simmer, cover the pan tightly and cook for another 10 minutes. Drain off the excess moisture and mash the beans well.

4. Add the cooked ingredients to the beans and mash again. Season with lots of black pepper (6–8 turns of the pepper mill) and serve.

Pesto Chicken and Asparagus

Serves 2

This tasty dish is really quick to make.

1 tbsp olive oil
2 small chicken breasts, chopped
6 large or 8 small spears fresh asparagus per person
200ml (7fl oz/ 3/4 cup) fromage frais
Small tin of chopped tomatoes, drained
1 tbsp green pesto

1. Heat the olive oil in a saucepan and add the chopped chicken. Cook gently for 10 minutes.

2. Steam the asparagus gently for 4–5 minutes or until tender.

3. Mix together the fromage frais, drained tomatoes and pesto. Add to the chicken, mix well and serve with the asparagus.

Vegetarian Moussaka

Serves 4

455g (1lb) aubergine (eggplant), sliced
5 tbsp olive oil
225g (8oz) onion, chopped
1 clove of garlic, crushed
225g (8oz) mushrooms, sliced
255g (9oz) plum tomatoes, drained, reserving the juice
285g (10oz) cooked red lentils
1 tsp dried mixed herbs
250ml (9fl oz/1 1/4 cups) water or vegetable stock
140g (5oz) cheese, grated

1. Flash-fry the aubergine (eggplant) in 4 tbsp of olive oil, then remove and drain.
2. Add the remaining olive oil and toss the onion and garlic for 1–2 minutes.
3. Add the mushrooms, drained plum tomatoes, cooked lentils, herbs and water/stock. Bring to the boil, cover and simmer for 5–10 minutes. If the mixture looks too dry, add the reserved tomato juice.
4. In a casserole dish, alternate layers of aubergine and the vegetable and lentil mixture. Sprinkle with grated cheese and bake for 20 minutes at 180°C/350°F/Gas Mark 4.

Fish and Fennel

Serves 4

Unsmoked haddock or cod are particularly good choices for this dish, which takes about 10 minutes to prepare and cook. Serve with green leafy vegetables of your choice.

> 2 cloves of garlic, crushed
> 1/2 cup fennel, finely chopped
> 4 fish fillets (select variety according to preference and best buy on the day)
> 1 tbsp olive oil
> 1 tbsp lemon juice

1. Combine the garlic and fennel in a small bowl.
2. Cut three narrow slits in each fish fillet and fill with the garlic and fennel mixture.
3. In a small bowl, mix together the oil and lemon juice. Brush one side of each fish fillet with this mixture.
4. Place the fish under a preheated grill with the oiled sides facing up and cook for about 3 minutes. Turn and brush the other side of the fillets with the oil/lemon mixture and return to the grill until cooked.
5. Fish is cooked when it is opaque in the thickest part but still moist – take care not to overcook.

Tofu Stir-fry

This delicious stir-fry takes about half an hour to prepare and cook, but is well worth it.

Serves 2

> 1/2 packet firm tofu
> 4 cloves of garlic, crushed
> 5cm (2 in) fresh ginger, grated
> Soya sauce

Cayenne pepper to taste

2 tbsp olive oil

225g (8oz) fresh vegetables to taste – broccoli, red onion, red/green/yellow (bell) peppers, mangetout, snow peas or any other vegetable of your choice

225g (8oz) bean sprouts

55g (2oz) slivered almonds

1. Cut the tofu into squares.
2. Make the marinade with the garlic, ginger, enough soya sauce to just about cover the tofu, and cayenne to taste. Marinate the tofu while you prepare the vegetables, cutting them into bite-sized pieces. (For a better taste, marinate the tofu overnight or for at least a couple of hours.)
3. Heat the oil in a wok or large saucepan. When hot, add the tofu and marinade. Stir-fry until the tofu is heated through.
4. Add the vegetables and bean sprouts and stir-fry until lightly cooked. Add the slivered almonds when serving.

Salad Niçoise

Serves 2 as a main course or 4 as a starter

8 tiny new potatoes

2 organic eggs

2 anchovies

2 tbsp olive oil

115g (4oz) fresh tuna

115g (4oz) green beans

1 tbsp balsamic vinegar

8 Greek olives

8 tiny yellow and red miniature tomatoes

Fresh parsley to taste

1. Boil the new potatoes in salted water until cooked.
2. Hard-boil the eggs, then peel and slice into segments.
3. Add 1 tbsp of the oil to a frying pan, sear the tuna, then cook on medium-high heat for a few minutes until cooked all the way through. Cool and slice into strips or small chunks.
4. Steam the green beans for only 2–3 minutes so they keep their bright green colour and stay crisp.
5. Mix together the balsamic vinegar and the remaining olive oil.
6. Mix all the ingredients together and serve.

Lunchtime Tortilla Wrap

Serves 1

2 oz chicken or turkey
2 lettuce leaves
1/2 ripe avocado
1 tomato
1 tbs '3&6 Mix'
1 tbs Essential Balance

1 Take one tortilla and use straight from the packet or warm according to manufacturers' instructions.
2 Place lettuce leaves over tortilla and arrange sliced avocado and tomato on top.
3 Sprinkle with '3&6 Mix' and Essential Balance.
4 Place chicken/turkey over mixture and wrap carefully.

Chicken Sautéed with Pears and Pine Nuts

Serves 4

This is an interesting and unusual way to serve chicken.

4 small chicken breasts, cut into pieces
Sea salt and black pepper to taste
Flour for dredging
1 tbsp butter
1 tbsp vegetable oil
1 leek, white part only, thinly sliced
3 tbsp currants
2 firm pears (Bartlett or Comice), peeled, cored and cut into thin wedges
1/4 cup balsamic vinegar
1 cup chicken stock
1/4 cup toasted pine nuts

1. Sprinkle the chicken with salt and pepper and dredge in flour, shaking off excess.
2. In a large skillet, heat the butter and oil. Cook the chicken on all sides until golden then remove from the pan.
3. Add the leek and currants to the skillet and cook for 5 minutes. Add the pears and cook for 2 minutes. Stir in the vinegar and stock, bring to the boil and cook for about 3 minutes, until syrupy.
4. Return the chicken to the skillet, cover and simmer for 20 minutes. Taste for seasoning and serve sprinkled with pine nuts.

Watercress and Endive Salad with Pears and Cheese

Serves 4

A young Gorgonzola cheese goes well with this colourful and tasty salad.

    2 heads Belgian endive, cut into thin strips
    2 bunches watercress, coarse stems removed
    2 ripe pears, cored and cubed
    1/4 cup white wine vinegar
    1 tsp Dijon mustard
    1 tbsp parsley, chopped
    1/2 cup olive oil
    115g (4oz) sweet cheese, crumbled

1. In a salad bowl, combine the endive, watercress and pears.
2. In another bowl, whisk the vinegar, mustard, parsley and olive oil until blended.
3. Toss the salad with the dressing and sprinkle with cheese of your choice.

Vegetable Dishes

Nearly all vegetables are alkaline forming, and adding them to an acid-forming meal is an easy way to make it more balanced. Rich in vitamins and low in calories, they are also some of the easiest foods to digest when properly prepared. They combine well with most foods, including proteins, grains and fats.

Vegetables have a wide range of tastes, from sweet (winter squash), bitter (dark-green leafy vegetables) and astringent (asparagus) to pungent (watercress) and salty (celery). This means they can be used in a variety of ways, both in cooking and in healing.

Steamed Asparagus

Serves 4

Asparagus ranks highly in the pH Diet. It is used medicinally in Ayurvedic medicine as a mild laxative, cardiac and nerve sedative, tonic and aphrodisiac.

> 1 cup water
> 455g (1lb) fresh asparagus
> 1 tbsp ghee or butter

1. Put the water in a saucepan, with a steaming attachment and place over a medium heat.
2. Wash and trim the asparagus, cutting off any woody ends (these can be saved for use in a soup stock if you like). Leave the asparagus stalks whole.
3. When the water has come to a boil, put the asparagus in the steamer and cover. Steam for 5–8 minutes until tender (a fork will go through them easily when ready).
4. Drain and serve with ghee or butter.

Green Beans and Almonds

Serves 2

225g (8oz) green beans
30g (1oz) slivered almonds

1. Steam the green beans until lightly cooked.
2. Sprinkle with the slivered almonds just before serving.

Sautéed Okra

Serves 4

455g (1lb) fresh okra
4 tbsp sunflower oil or ghee
1 tsp fenugreek seeds
$1/2$ tsp turmeric
$1/2$ tsp sea salt
2 tsp coriander (cilantro) powder
$1/4$ tsp black pepper
$1/4$ tsp ground cumin

1. Wash the okra and spread it on absorbent paper towel or cloth to dry thoroughly. Cut into 1cm ($1/2$-inch) pieces, throwing away the tops.
2. In a heavy skillet, heat the oil or ghee and add the fenugreek seeds. When they turn brown, add the turmeric, salt and okra. Mix well, cover and cook over a low heat for 10 minutes.
3. Uncover and cook for another 10 minutes over a very low heat, stirring occasionally to keep the okra from sticking.
4. Add the remaining ingredients and mix well. Serve with a main meal.

Roasted Vegetables

You can add any vegetables you like to this delicious dish. It's fun to experiment. A few suggestions are listed here.

Vegetables of your choice, such as green beans, red (bell) peppers, red onions, chopped
Olive oil
Garlic, crushed
Sea salt and freshly ground black pepper

1. Prepare as many vegetables as you need and combine with olive oil, garlic and seasonings to taste.
2. Pile into an ovenproof dish and roast at a high temperature until cooked (approximately 30 minutes, depending on quantity).

Winter Squash and Ginger

This is a really different vegetable dish.

6–8 cubes of winter squash per person
Olive oil
Fresh ginger, grated
'3&6 Mix'

1. Steam the cubes of winter squash.
2. Dress with the olive oil, fresh ginger and '3&6 Mix'.

Fresh Mango Salsa

Makes about 4 servings

This salsa is a delicious accompaniment to chicken, turkey or other protein foods.

    1 ripe mango
    1 jalapeño pepper
    2 tablespoons fresh coriander (cilantro)
    Zest and juice of 1 lime
    1 teaspoon extra virgin olive oil

1. Dice the mango and pepper and mix together.
2. Finely chop the coriander, mix with the lime zest and juice, add the olive oil and mix well.
3. Combine with the mango and pepper and refrigerate until ready to use.

Snacks

It is good practice to eat three meals a day at regular times, as eating between meals can overwhelm the digestive tract and sometimes ruin the appetite. Having said that, for people with blood sugar imbalances or erratic eating schedules, snacks can be invaluable. As with main meals, make wise choices with snacks. A snack does not mean chocolate biscuits or junk food, and you should only have a snack if you are genuinely hungry. Fruits make excellent snacks. Nuts and seeds are quick and easy, and Ayurvedic medicine regards most nuts as restorative, nutritious and warming in their action.

Toasted Sunflower Seeds

Makes 10 x 1 tbsp servings

This all-purpose snack food is rich in potassium and zinc. Sunflower seeds also make a tasty garnish in salads, vegetables and main dishes.

170g (6oz) fresh, raw sunflower seeds

1. Warm a large, heavy skillet over a low heat. After 2 minutes, add the sunflower seeds.
2. Toast for 15–20 minutes, stirring occasionally.
3. Cool and serve.

Dry-roasted Pumpkin Seeds

Makes 10 x 1 tbsp servings

The added spices make it easier to digest the seeds. The pumpkin seeds turn a lovely bright green when roasting.

170g (6oz) raw pumpkin seeds
1/2 tsp ground cumin
1 tsp coriander (cilantro) powder
1/4 tsp turmeric
1/2 tsp rock or sea salt

1. Mix all the ingredients in a large, dry skillet and cook over a low heat for about 10 minutes until the pumpkin seeds begin to pop.
2. Stir and cook for another 1–2 minutes. Cool.

Spiced-up Yoghurt

Serves 1

This is rich in calcium, iron, sulphur and B-vitamins, and can be calming for the nerves. Excellent as a snack or served with Bharti's Bread (*page 148*) for a satisfying lunch.

    1 cup plain yoghurt
    1 tsp blackstrap molasses
    1/4 tsp vanilla extract

1. Stir all the ingredients together and serve.

Tortilla Desert/Snack

    5 oz strawberries
    2 oz cherries (de-stoned)
    1 tbs '3&6 Mix'
    2 tbs thick yoghurt

1 Take one tortilla and use straight from the packet or warm according to manufacturers' instructions.
2 Spread the yoghurt over the tortilla and sprinkle the '3&6 Mix' over.
3 Arrange the strawberries and cherries on top, before wrapping. Delicious!

Juices: Cleansing and Healing

Fresh juices have remarkable cleansing and restorative powers. Vegetable juices restore vital mineral (alkaline) reserves to the body and are healing. As most vegetable juices are alkaline forming, they can be used to balance any meal where the protein level may be high. Fruit juices can be described more as cleansing as they remove toxins from the body. Juices are also natural anti-diuretics. Dilute with 50 per cent water and give to children to boost their nutrient intake. Juices can be invaluable for the elderly who may have difficulty eating hard fruits, and for people recuperating from illness.

Juices are a quick and easy means of boosting your energy levels, and a great way to get your 'five a day'. They are an important part of the pH Diet at both Level 2 and Level 3. Vegetable juices are used in Level 2 to build up the alkaline reserves, and Level 3 introduces fruit juices for detoxifying, cleansing and to give vitality.

Remember, it is very important to start with the vegetable juices and proceed to the fruit juices, not the other way around! After a couple of weeks on vegetable juices (or when your urine and saliva pH measures good alkaline reserves), you can mix vegetables with fruits, before taking the fruit juices alone. Fruit juices are mildly acidic and should only be taken when alkaline reserves have been built up. They can also cause a rapid rise in blood sugar, and anyone suffering from candidiasis should be cautious about excessive sugar intake.

Most of the recipes here are for either vegetable juices or fruit juices. A note of caution, when mixing vegetable and fruit juices in the same glass or you may well suffer from embarrassing flatulence! The exceptions are apple and carrot, which you can mix with anything.

Aim to juice immediately before drinking and on an empty stomach for optimal benefits. Always dilute dark-green vegetable juices (broccoli, spinach and watercress) and dark-red vegetable juices (beetroot, red cabbage) by four parts to one. They are very potent in taste and effect!

If you don't have time to make your own vegetable juices, you can use V8 juice or any prepared vegetable juice, but check the label for added sugars or preservatives. At Level 3, juice should ideally be prepared at home, immediately before serving, so that you get as many of the natural enzymes and nutrients as possible.

You will need a juicer for all these recipes. You keep the kettle on your work surface because you use it every day, and this is where the juicer needs to be.

Simple Vegetable Juices for Level 2

Green Power Juice

Serves 2

This is a power-packed green drink. Go easy on the parsley as it has a very strong flavour.

    2–3 stalks celery, chopped
    1 cucumber, chopped
    2–3 large kale leaves
    4–5 large lettuce leaves
    2 cups spinach
    1/4–1/2 cup fresh parsley

1. Wash all the ingredients well and put them through the juicer.
2. Drink immediately.

Mixed Vegetable Juice

Serves 2

Excellent for the nervous system and a powerful diuretic.

> 1 large carrot, chopped
> 3 stalks celery, chopped
> 1 head lettuce
> 1 bunch parsley
> 2 small courgettes (zucchini), chopped (optional)

1. Wash all the ingredients well and put them through the juicer.
2. Drink immediately.

Carrot and Ginger Juice

Serves 2

This is a tonic for the liver, and is especially useful if it is sluggish or has been abused by alcohol. Best drunk on an empty stomach.

> 2.5–5cm (1–2 in) fresh ginger root, chopped
> 8 large carrots, chopped

1. Peel the ginger and wash the carrots.
2. Put the carrots and ginger through the juicer.
3. Drink immediately if possible.

Carrot and Parsley Juice

Serves 2

> 6 carrots, chopped
> 1 handful of parsley

1. Blend together the ingredients.
2. Drink immediately.

Liver Cleanser

Serves 1

> 2–3 carrots, chopped
> 1/2 raw beetroot, chopped

1. Blend together the ingredients.
2. Drink immediately.

Body Cleanser

> 4 carrots, chopped
> 1/2 cucumber, chopped
> 1 raw beetroot, chopped

1. Blend together the ingredients.
2. Drink immediately.

Potassium Special

Serves 2

This is an excellent juice for replenishing potassium at Level 2.

    1 handful spinach
    1 handful parsley
    2 stalks celery, chopped
    4–6 carrots, chopped

1. Blend together the ingredients.
2. Drink immediately.

Alkaline Special

Serves 1

Cabbage, an alkaline super food, is full of minerals that also protect against colon cancer.

    1/4 head cabbage (any variety), chopped
    3 stalks celery, chopped

1. Blend together the ingredients.
2. Drink immediately.

Nail Strengthener

Serves 1

115g (4oz) bean sprouts
115g (4oz) chuck white cabbage, chopped
2 large carrots, chopped

1. Blend together the ingredients.
2. Drink immediately.

Super Skin Vegetable Juice

Serves 1

1/4 red (bell) pepper, sliced
1/4 green (bell) pepper, sliced
1/3 large cucumber, chopped

1. Blend together the ingredients.
2. Drink immediately.

Simple Fruit Juices for Level 3

Colon Cleanser

Serves 1

This juice is full of fibre, which helps remove toxins – including cholesterol – from the body.

2 apples, chopped
1 pear, chopped

1. Blend together the ingredients.
2. Drink immediately.

Watermelon Juice

Serves 1

A refreshing, delicious and very cleansing juice.

1 large chunk of watermelon

1. Blend the watermelon chunk with the rind and seeds.
2. Drink immediately.

Summer Cocktail

Serves 1

2 apples, chopped
4–6 strawberries

1. Blend together the ingredients.
2. Drink immediately.

Melon Juice

Serves 1

Cantaloupe is the most alkaline-forming melon. Melon juice is best drunk without food and is especially beneficial mid-morning or mid-afternoon.

1/2 cantaloupe melon

1. Wash well before cutting into strips and juicing the melon, rind *and* seeds.
2. Drink immediately.

Carrot and Apple

Serves 1

This sweet juice is a real favourite.

    3 carrots, chopped
    1 apple, chopped

1. Blend together the ingredients.
2. Drink immediately.

The Waldorf

Serves 1

    1 stalk celery, chopped
    2 apples, chopped

1. Blend together the ingredients.
2. Drink immediately.

Super Skin Fruit Juice

Serves 1

85g (3oz) cranberries
1 1/2 apples, chopped

1. Blend together the ingredients.
2. Drink immediately.

Wrinkle Smoother

Serves 1

1 peach
140g (5oz) strawberries or raspberries
3 guavas

1. Blend together the ingredients.
2. Drink immediately.

Acne Cleanser

Serves 1

3 apricots
170g (6oz) cherries
1 nectarine

1. Blend together the ingredients.
2. Drink immediately.

Bone Builder

Serves 1

> 2 passion fruit, peeled
> 225g (8oz) grapes

1. Blend together the ingredients.
2. Drink immediately.

Breads

Bread is a staple of the British diet and it would be unrealistic to advise you never to eat bread again. Bread provides us with energy and a feeling of satiety, both of which are important. However, breads are also acid forming. If you want to eat bread, it is recommended that you either make your own, or buy bread from bakers who use organic ingredients. Try to look for yeast-free, unleavened and sprouted breads. Burgen bread comes highly recommended as it contains healthy ingredients, and is available in most supermarkets. Be sparing with the bread as one slice often leads to another! Don't forget tortillas – they are on the acceptable acid-forming food list and can be made into fajitas, quesadillas or enchiladas. One is enough for a good breakfast or lunch.

Bharti's Bread

Makes 1 loaf (7 slices)

This wonderful 'bread' is bursting with flavour. With the protein content from the quark, it makes a meal in itself. An excellent accompaniment to alkaline-forming soups and salads, it is also great for lunch boxes. It keeps well in the refrigerator.

2 tbsp corn oil

1 tbsp black mustard seeds (not yellow mustard seeds as these don't pop)

1 1/2 tsp sea salt

1/2 tsp freshly ground black pepper

1 small chilli pepper, seeded and chopped

1 large onion, chopped

1 large red (bell) pepper, seeded and chopped

140g (5oz/1 cup) gram flour

230ml (8fl oz/1 cup) soya milk (unsweetened)

3 medium organic eggs, beaten

225g (8oz/1 cup) quark, or other very low-fat soft cheese

1 tbsp garam masala (or commercial curry powder)

1. Preheat the oven to 170°C/325°F/Gas Mark 3.
2. Butter a 23cm (9in) square tin or lasagne dish.
3. Combine the oil, mustard seeds, salt, black pepper and chilli in a heavy frying pan (skillet) and heat until the seeds pop. Add the onion and red pepper, cover and cook until the vegetables are tender.
4. Meanwhile, combine the gram flour and soya milk in a large mixing bowl. Add the eggs gradually. (Gram flour is surprisingly heavy, so the mixture will be thick.)
5. Add the low-fat cheese and mix well to combine.
6. Stir in the cooked vegetables, spices and oil and the garam masala. Pour the mixture into the prepared dish and bake for 30 minutes, or until the centre of the bread is firm.

Corn Bread

Makes 1 loaf

> 170g (6oz/1½ cups) maize flour
> 170g (6oz/1½ cups) rice flour
> 2 tbsp '3&6 Mix'
> Pinch of sea salt
> 1 tbsp mixed herbs
> 1 tsp sodium or potassium bicarbonate
> 285ml (½ pint/1⅓ cups) milk (rice, goat's, soya)

1. Preheat the oven to 180°C/350°F/Gas Mark 4. Grease and line a 455g (1lb) loaf tin.
2. Mix together the dry ingredients in a large bowl. Make a well in the middle and pour in half the milk. Stir together and keep adding milk until the ingredients stick together to form a single lump (you may want to use your hands for this). You may need slightly less or more milk than given in the ingredients list.
3. Transfer the mixture to the loaf tin and bake for approximately 45 minutes. To test when cooked – insert a metal knife or skewer into the centre. If it comes out clean, the bread is ready, otherwise leave for a further 5 minutes. When ready, remove from the oven. If necessary, loosen the edges with a spatula or fish slice and tip out of the tin to cool, preferably onto a wire rack.

CHAPTER FOUR

How to Test Your pH Levels

Although testing your pH level is not essential when following the pH Diet, it is very useful. When you start monitoring and recording it on a regular basis, you become an active participant on your journey to health and beauty, rather than a passive passenger.

You will need a roll of special paper for testing your saliva and urine (*see Resources*). To get the most accurate results, we recommend you do these tests before you start Level 1 of the programme. The results will give you valuable information about how your body is handling the different types of foods over the course of the first few days. Instructions for carrying out these tests are given below (*see 'Urine Testing' and 'Saliva Testing'*).

Using special strips to test alkaline or acid status is quick and easy. Reading the results is easy too, but interpreting them can be a little tricky, as there are many factors to be taken into consideration.

Urine pH can range from 4.5 to 8.5. An ideal reading, after a balanced meal including protein, carbohydrates and fats, would be 6.1. This indicates that the body has sufficient alkaline reserves to deal with the acid wastes from the meal. Urine pH readings give the most accurate indication of the body's ability to handle food that has been consumed within the past 12 hours. They can also help in the assessment of a person's overall health.

The results of urine testing indicate how well your body is assimilating minerals – especially calcium, magnesium, sodium, potassium and zinc. These are called the 'acid buffers' because they control the acid level in the body. If there is too much acid, the body will not be able to excrete it. It will either store it in body tissues (autointoxication) or buffer it from minerals absorbed from vegetables and fruit in the diet. When insufficient absorbed minerals are available, the body will 'borrow' the minerals from organs and bones in order to neutralize the acidity.

When the body excretes urine, it is getting rid of substances it cannot use or that are harmful. The kidneys filter out any beneficial substances the body can use, such as glucose and mineral salts, and diffuse them back into the bloodstream. Harmful substances are passed down the urethra tubes to be excreted as urine. If your diet contains lots of vegetables then natural waste products will be eliminated in the urine. People on a high-protein diet or diet high in acid-forming foods and drinks will excrete ammonia, bicarbonate and perhaps protein. The body does its best to rid itself of waste from recent meals, snacks, drinks and stimulants.

An ideal saliva pH upon waking is 6.8. The results of the saliva tests show the efficiency of the liver, stomach and digestive system.

Urine Testing

To test your urine, tear a small strip (about 7cm/3in) from the roll of special paper. Simply hold the strip of paper beneath the flow of urine, making sure it is wet through. Alternatively, you can collect a small amount of urine in a disposable container and dip the strip into it. With both methods, compare the colour of the strip to the chart on the dispenser within a few seconds of wetting it. Make a note of the reading.

Day 1: No Specific Meal Test

On the first day, eat and drink exactly the same as you normally do. Make absolutely no change at this stage.

Day 2: First Urine Test

Test the first morning urine with the test strip and keep a record of the result. This will reflect your body's ability to metabolize and neutralize the acid wastes from your current diet – the food and drinks you had on Day 1.

Day 2: Your Normal Diet + Acid Meal Test

For the second test, eat and drink normally but have acid-forming foods *only* as an evening meal. Acid-forming foods are usually high in protein. Suitable meal choices would be:

- bacon, sausages, ham, burgers, pork chops, steak (or any beef, lamb or pork)
- potatoes – boiled, chipped, mashed
- pasta with a meat sauce and Parmesan or other cheese
- white rice with meat or commercially made sauces (curry, chilli con carne)
- bread or rolls with margarine or butter
- cheese and biscuits
- biscuits, cakes, ice cream or other dessert
- alcohol, milk, milkshakes, coffee, tea with milk and sugar

You could also have eggs, chicken or turkey, as these are all acid-forming protein foods, although they are of a much better quality and not as taxing on the body. For accurate results, we really want to 'challenge' the body.

Make sure you eat no fruit or vegetables from 2pm. Accompany your meal with a glass or two of wine if you choose. Drink tea and coffee as normal.

Day 3: Second Urine Test

The next morning (or after eight hours' sleep), check your urine pH measurement with a testing strip and note the result. As the meal the previous evening was very acid forming, the body will need to eliminate this acid from the body. The test result should reflect this in a lower pH measurement.

Day 3: Your Normal Diet + Alkaline Meal Test

For the third test, eat your normal diet with the addition of an alkalizing evening meal of basically **all vegetables**. You can choose any amount of green leafy vegetables, plus broccoli, green beans, asparagus, avocado and tomatoes – anything from the list of alkaline-forming foods on pages 88–9 – **but no proteins, grains or fats after 2pm**. The earlier you start the challenge, the more accurate the result will be.

Day 4: Third Urine Test

The next morning (or after eight hours' sleep), check your urine pH measurement with a testing strip and note the result. If you cheat the evening before and have *anything* acid forming (alcohol, coffee, protein foods, pasta), the results will not be accurate.

The minimum number of times you should measure your pH is shown in the chart below. However, you can test as often as you like – every day, once a week or three times a week, if you so wish – to monitor your progress more closely. Urine readings change from day to day and are a good indicator of how your body responds to the stresses of the previous day. Eating too many of the wrong foods will show in your readings on the morning of the following day.

Use this chart to record the results of your tests.

URINE pH	NON-SPECIFIC MEAL	ACID MEAL	ALKALINE MEAL
Week 1 – before starting Level 1 of programme Date:			
After 1 month on Level 1 Date:			
After 1 month (or when alkaline reserves have been built up) on Level 2 Date:			
After 1 month on Level 3 Date:			

Interpreting the Results

Day 2: First Urine Test
The ideal pH of your urine after eating just your normal dietary choices would be 6.1.

Day 3: Second Urine Test
The best result at this stage would be a pH between 4.5 and 6.0. This would indicate that your body has sufficient alkaline reserves to buffer the acids, and that your kidneys and adrenal glands have enough energy to dispose of them.

If, after the acid-forming evening meal, your urine pH is 6.8 or higher, this is not good. This indicates that the body's mineral reserves are exhausted.

If your urine pH is between 6.0 and 6.6, this indicates that the body's mineral reserves are reduced. The higher the pH, the worse the situation.

Poor results mean you need to take definite remedial action to build up your alkaline reserves. Work hard on Level 1 and remain on Level 2 until your reserves have been restored. You may want to consider making an appointment with a nutritional therapist who will interpret the pH results, advise you on the next course of action and probably recommend some supplements.

Day 4: Third Urine Test
If the pH range is 6.8–8.5, there could be two possible explanations. It probably means all is well *if* you are perfectly healthy and free of symptoms. However, if you are experiencing symptoms of ill-health, this alkaline response to an alkaline meal could be an indication that your cells are too toxic to use any alkaline reserves and are being dumped instead. If you have symptoms of ill-health with an alkaline pH response to an alkaline challenge, you need to work hard at Level 1 of the programme to reduce your toxic load, and stay on

Level 2 until your urine reaches an ideal pH of 6.1.

If you are genuinely eating a well-balanced diet, containing lots of alkaline-forming vegetables, and your urine pH is constantly high (alkaline reading), then you may want to consider making an appointment with a nutritional therapist who will interpret the pH readings, advise on the next course of action and probably recommend some supplements.

If your pH ranges from 5.5 to 6.1 after an alkaline-forming meal, you may have a better level of alkaline reserves.

If the pH range is 4.5–5.5, it means your body has excess stores of acid. You need to stay on Level 2 of the programme until the result is higher.

All these results are summarized in the following tables.

Evening Meal 6-7pm	pH of urine sample taken first thing upon waking	Result	Correction
Acid-forming foods: meal high in meat, fish, cheese, eggs, nuts and grains.	6.8–8.5 Not a good reading.	Alkaline reserves exhausted. Probably means depleted alkaline reserves and possible exhausted adrenal glands as well as probable digestive problems.	Start at Level 1: Reducing the Toxic Load. Do not make sudden changes to your diet. Be cautious and patient.
	6.0–6.6 Not so good.	The body is trying hard to eliminate acidic wastes. The higher the pH, the worse the situation. Alkaline reserves are very low and the body is struggling to produce an acid urine.	Start at Level 1: Reducing the Toxis Load. Reduce acid-forming foods, drinks and stimulants from the diet and very gradually increase servings of vegetables.

Implications of Urine pH Readings after Acid-forming Meal Test

Evening Meal 6-7pm	pH of urine sample taken first thing upon waking	Result	Correction
	4.5–6.0 A healthy result.	Too much dietary protein so the body is losing sodium fast, but alkaline reserves still available to buffer the acid. Also indicates the adrenals and kidneys have appropriate energy to dispose of accumulated acid deposits.	Start at Level 1: Reducing the Toxic Load. A good result so you can spend less time on Level 1 before moving on. Reduce acid-forming foods in diet and gradually increase servings of vegetables then fruit to maintain the good result and prevent future imbalance.

Evening Meal 6-7pm	pH of urine sample taken first thing upon waking	Result	Correction
Alkaline-forming foods: meal high in vegetables and foods from list of alkaline-forming foods (*pages 88–9*).	6.8–8.5 Two possible explanations.	Either: 1. Everything is normal, you are very healthy and have *no* symptoms of ill-health *or* 2. Cells are so toxic they cannot use alkaline minerals from foods eaten. *Depends on how you feel.*	1. If you feel healthy and have **no** symptoms of ill-health, you can **work through the first two levels speedily or go straight to Level 3**. 2. You are likely to suffer from many symptoms of ill-health. Start at Level 1: Reducing the Toxic Load. You may also need to consider taking nutritional supplements.
	5.5–6.1 Quite a good reading.	A better level of alkaline reserves, but the key is how you actually feel and if you have any symptoms. If you feel healthy, this range is acceptable.	Start at Level 1: Reducing the Toxic Load and proceed quickly to Level 2. Alkaline reserves need to be restored, indicated by a consistent urine pH reading of 6.1.

Implications of Urine pH Readings after Alkaline-forming Meal Test

Evening Meal 6-7pm	pH of urine sample taken first thing upon waking	Result	Correction
	4.5–5.5 Not a good response.	This means your body has excess acidity stored. The organic minerals from alkaline foods are going directly to cells. No alkaline reserves have been built up.	Start at Level 1: Reducing the Toxic Load. At Level 2, pH will rise slowly with the increase in vegetables and when the demands of cells have been satisfied.

Saliva Testing

You may also want to test the pH of your saliva. The results of saliva testing indicate the activity of digestive enzymes in the body, especially in the liver and stomach.

Digestion starts when we see or smell food. The saliva that builds up in the mouth contains amylase, an enzyme responsible for breaking down carbohydrates. To function properly, this enzyme needs a pH of around 6.8. If you have adequate alkaline reserves in your body, testing your saliva pH as you salivate before a meal should give you a pH reading of around 6.8. If your pH is not getting up to at least 6.8, you can assume there is stress in your alkaline reserves.

The further below 7 it goes, the more depleted are these reserves. A low reading also indicates that the digestive system generally is not doing so well. You will need to spend longer on Level 2 of the programme to restore alkaline reserves.

Day 4: Saliva Test

After carrying out the urine pH test first thing in the morning, test your saliva before *anything* goes into your mouth. Compare the colour of the strip to the chart on the dispenser *within a few seconds* of wetting it. Make a note of the readings.

For optimum health, saliva should be 6.8 or thereabouts – a dark-green colour on the strip. After the test, eat your usual breakfast. When finished, wait for two minutes and test again, using a fresh strip. The strip should now be dark blue, indicating a pH of 8.0. After another two minutes, test again, when the reading should be about 7.5. After just two more minutes, take a final reading. The result should be 6.8, the same as at the beginning.

Ideal Saliva pH Readings

1st test (immediately upon waking)	6.8
2nd test (2 minutes after eating breakfast)	8.0
3rd test (after 4 minutes)	7.5
4th test (after 6 minutes)	6.8

If these are your readings, your body is in an optimal state of health and your regulatory mechanisms are working efficiently.

If your readings are consistently lower or do not change after eating, you are one of many millions of people whose readings are not ideal. By making a few well-controlled improvements to your diet, you could radically improve your general health – starting with Level 1.

You can test your saliva as often as you want, but we recommend you make a note of it on a weekly basis for the first four weeks, and then once a month. This information will be helpful should you decide at any time to visit a nutritional therapist. Use the following charts to record your results.

SALIVA pH	1st Reading	Week 1	Week 2	Week 3	Week 4
1st test before food					
2nd test 2 minutes after					
3rd test 4 minutes after					
4th test 6 minutes after					

SALIVA pH	1 month	2 months	3 months	4 months	5 months
1st test before food					
2nd test 2 minutes after					
3rd test 4 minutes after					
4th test 6 minutes after					

CHAPTER FIVE

The Benefits of the pH Diet

As there is not enough room in this book to describe the countless benefits of the pH Diet, we will look at the five most important:

- preventing and reversing osteoporosis
- looking great
- managing your weight
- fighting fatigue
- reducing the symptoms of arthritis

Preventing and Reversing Osteoporosis

The health of our bones, like every cell, tissue and organ in the body, is affected by our diet. Bone, our nutrient 'bank' of minerals, is just like any other bank: the more deposits and the fewer withdrawals you make now, the more you will have later in life, and it is never too late to start depositing.

At the age of 35, a woman's battle against bone loss begins, and it intensifies with menopause and beyond. A correct diet can substantially increase a woman's chance of maintaining and even regaining normal bone mass. The ideal diet for bone health is the pH Diet.

What is Osteoporosis?

Osteoporosis causes brittle, thin and porous bones. It is sometimes described as 'the silent epidemic', as people often only discover they have it when they break a bone. We sometimes forget that bone is an active, living tissue, continually breaking down old, damaged bone and building up new bone. This 'remodelling' process occurs in all bones throughout life and serves two purposes. First, it helps keep bones 'young' by replacing old or weakened areas with new, well-formed tissue. The second function is to make the bones better able to meet the everyday demands placed on them.

The bone matrix consists of an organic component – primarily collagen and other proteins – and an inorganic component, which is responsible for the rigidity of bone. The inorganic component is composed mainly of calcium phosphate and calcium carbonate, with small amounts of magnesium, fluoride, sulphate and other trace minerals.

The risk of developing fractures is definitely related to bone mineral content, known as 'bone mass'. However, thinning of the bones alone may not be sufficient to result in a fracture. Diet can be a vital factor.

What You Eat Affects Your Bones

The alkaline-forming foods we recommend in *The pH Diet* supply all the calcium, potassium, magnesium and other important minerals for optimal bone growth and repair. Limiting your intake of acid-forming foods to 20–30 per cent of your diet prevents calcium and other minerals being lost from the bones.

Vitamin K also plays a critical role in maintaining healthy bones and preventing osteoporosis, and following the pH Diet will provide you with ample amounts. It was once thought that the only role of vitamin K was in blood clotting, but new research has clearly shown the part played by vitamin K in bone metabolism. You do not need to take a supplement if you eat enough vegetables containing vitamin K, namely vegetables from the cruciferous family like broccoli and cauliflower.

Many nutritionists believe that degenerative diseases like osteoporosis are caused, at least in part, by our modern diet. Too much sugar, fat, salt, refined flour, caffeine, alcohol, processed foods and food additives in the diet can lead to a build up of acid in the body. Each level of the pH Diet 'cleans up' the diet by reducing the consumption of anti-nutrients and stimulants, and by increasing the intake of water, whole grains, fruits, vegetables, nuts and seeds, beans and other unprocessed foods. Although it is impossible to determine the precise effect of diet on bone health, there is some evidence that the typical Western diet promotes the development of osteoporosis.

The Role of pH in Osteoporosis

The role of pH in osteoporosis is complex, but here is a simplified explanation of how calcium loss occurs in the bones.

As part of normal metabolism, the body produces acids. If there is too much acid or too few alkaline minerals to neutralize it, the body cannot eliminate it. The body then stores the acid, mostly in the spaces around the cells (interstitial spaces). The intention is that the acid will eventually be removed. The body knows that for every molecule of acid that gets stored in the tissues, an equal molecule of alkaline substance needs to be put into the blood because one day it will need to escort the acid out of the body. This is the body's amazing compensatory mechanism at work.

What we see here is the pH interplay between the blood and the tissues. If the body has an acid overload, it stores the acid in the tissues (the tissues' pH decreases). The blood compensates and becomes alkaline (the blood's pH increases).

If acid stores build up, the body's health will deteriorate. The blood pH will often turn from its overly alkaline condition and start to move down the pH scale to become more acidic. When that happens, the body is losing its compensatory mechanisms. If this dangerous situation is not immediately addressed, it can lead to death. Don't panic! Before that happens, the spaces around the cells become saturated as more acid accumulates. As a result, acid wastes can end up *inside* the cell, rather than in the spaces around the cells. When acid gets pushed into the cell, it displaces potassium and magnesium, then sodium. These are three critical minerals in our body. The potassium and magnesium will leave the body but, as a preservation mechanism, the sodium will be retained.

Now remember, the body knows it must place an alkaline molecule in the blood to escort out the increasing acid that is being stored in the tissues and cells. What it will often do when mineral reserves are low is draw calcium from the bones and put it into the blood. This plays a major role in osteoporosis and arthritic pain.

What you should not do in this case is take more calcium supplements. It is important to understand that osteoporosis is *not* caused by calcium deficiency. It is the result of too many acid-producing foods in the diet, forcing the body to use available calcium from the bones to neutralize it. In these situations, what you really need is more magnesium, potassium and zinc and perhaps organic sodium, which all help to break down acids.

Calcium

The mineral calcium plays a vital role in health and in the pH balance of the body. It is the body's most abundant mineral with 99 per cent found in the bones and the other vital 1 per cent in the bloodstream. When blood calcium levels drop, calcium elements are pulled from the bones to bring the blood calcium back in balance. When the diet is deficient in calcium, or calcium is not being absorbed, the calcium from bone will not be replaced.

Symptoms of calcium deficiency include muscle cramps, tremors or spasms, insomnia, nervousness, joint pain, osteoarthritis, tooth decay and high blood pressure. Severe deficiency causes osteoporosis. As we have seen, however, this is more likely to be connected to excess protein and hormone imbalances.

Numerous studies have shown that calcium supplementation at levels of 1,000–1,500mg a day can help reduce bone loss by 30–50 per cent. However, not all calcium supplements are the same. Some are much easier than others for the body to absorb and use. Calcium supplements can also be expensive.

The body is better able to absorb calcium from certain foods than from others. The absorption of calcium from kale, for example, is 41 per cent compared to a 32 per cent from milk. One cup of most dark-green leafy vegetables contains nearly as much calcium as a cup of milk. Sea vegetables actually have more. It is also thought that the protein in dairy products may inhibit calcium absorption, and even increase calcium loss.

Dairy-free sources of calcium

FOOD	AMOUNT	CALCIUM (mg)
Wakame (sea vegetable)	½ cup	1,700
Agar agar (sea vegetable)	½ cup	1,000
Sardines with bones	½ cup	500
Sesame seeds	¼ cup	500
Tinned red salmon	½ cup	275
Soya yoghurt	1 cup	272
Broccoli	1 cup	178
Almonds	¼ cup	175
Tofu	1 cup	150
Soya beans	1 cup	130
Sunflower seeds	¼ cup	70
Lima beans	1 cup	60
Lentils	1 cup	50
Romaine lettuce	1 cup	40
Mung bean sprouts	1 cup	35
Alfalfa sprouts	1 cup	25

Wherever your calcium comes from, it won't do its work left to its own devices. In order to be absorbed, it must have the cooperation of a number of supportive nutrients. Vitamin D and magnesium are needed to maintain adequate levels of calcium in the blood, thereby reducing the chances of the body stealing reserves from bones. Calcium also maintains a delicate balancing act with the mineral phosphorus: when phosphorus levels are excessive, calcium will be excreted. (Phosphorus is high in fizzy drinks, and adolescent girls in particular should be discouraged from drinking them.)

Calcium supplements should not be necessary if you eat plenty of foods rich in calcium and magnesium (dairy-free sources of calcium and dark-green leafy vegetables). However, should you decide that you do need calcium supplements, always choose a balanced supplement containing calcium, magnesium and vitamin D (see Resources).

Hormone Replacement Therapy

One of the most difficult decisions faced by women entering menopause is whether or not to take hormone replacement therapy (HRT). This form of oestrogen treatment has obvious benefits, such as the relief of hot flushes, depression and vaginal atrophy. Oestrogen has also been clearly shown to slow the rate of post-menopausal bone loss and to reduce the incidence of osteoporotic fractures by about 50 per cent. However, there are also definite risks and side-effects associated with taking synthetic oestrogen.

Studies show that women taking HRT are between four and thirteen times more likely to develop cancer of the uterus than women who are not taking oestrogen. This risk can be reduced if the HRT includes the hormone progesterone as well as oestrogen. Many studies have proved beyond doubt that synthetic oestrogens promote the growth of oestrogen-sensitive breast cancers.

Some women really benefit from HRT, but if you are considering taking synthetic hormones into your body, read everything you can on the topic. There is new evidence every week suggesting more alternative and natural ways of beating menopause symptoms, ways that do not increase your risk of uterine or breast cancers. Do not take your doctor's word that it is the best thing for you – you are responsible for your own body and what you put into it. The pH Diet includes foods containing a natural form of oestrogen, called phyt-oestrogens.

How Does Oestrogen Prevent Osteoporosis?
The fact that osteoporosis is far more common in women than in men and that bone loss accelerates after menopause suggests that an age-related decline in female sex hormones plays an important role in the development of osteoporosis. This concept is supported by the observation that pre-menopausal women whose ovaries have been surgically removed lose bone at an unusually rapid rate for about four to six years following the operation. In women with intact ovaries, the amount of oestrogen and other hormones secreted by the ovaries begins to decline around the time of menopause. The adrenal glands compensate in part for this by secreting certain androgens (male hormones) into the bloodstream, which are converted into oestrogens elsewhere in the body. Despite this contribution from the adrenal glands, however, the amount of oestrogen in the body falls significantly at menopause.

HRT works by preventing old bone being broken down. In other words, when taking HRT, the condition of the bones should remain the same as when you started the therapy. At the same time, however, HRT prevents new bone being made, so it cannot reverse established osteoporosis. When HRT is discontinued, bone loss resumes, possibly at an accelerated rate. Therefore, for HRT to be successful in the prevention of osteoporosis, it must be started very early, before significant bone loss has occurred, and continued indefinitely.

Osteoporosis Risk-assessment Test

There is a urine test you can do at home to assess your risk of developing osteoporosis. This is called the Bone Reabsorption Test. It only involves taking a urine sample at home and sending it off to a laboratory for analysis. The result will be minimal risk, medium risk or high risk. If you want to take this test, you will need to do so through a clinical nutritionist (*see Resources*), as the results are returned to them for interpretation and recommendations. As prevention is better than cure, it is better to know if you are at risk as soon as possible so you can do something about it.

Looking Great

The appearance of your skin, hair, nails and eyes is one of the first indications of not only your age, but also your general health. As the pH Diet is bursting with vitamins, minerals and antioxidants, which significantly improve your general health, the condition of the skin, nails, hair and eyes will improve too. By nourishing your inner body with alkaline-forming super foods, your inner glow will naturally shine through into your outer appearance.

Healthy Skin

The Structure of the Skin
The largest organ in the body, the skin is your primary defence against the outside world. The skin is constantly growing. Skin cells are continually being made to renew the outer layers of the skin. Unfortunately, as you get older, the rate of new skin cells being made can drop by up to 50 per cent!

The skin is made up of two main parts:

The epidermis, the part you can actually touch, is exposed to the outside world. It is made up of dead skin cells that act as a water-proof material around us. A protein substance called keratin sur-rounds these dead skin cells, preventing them from flaking off. Layers of skin cells within the epidermis are constantly moving up to the outside, as the outer layer becomes sloughed off at the surface by wear and tear. The epidermis also contains a number of cells called melanocytes. These release melanin, the compound respon-sible for tanning and protecting the skin against UV radiation. As melanocytes also decrease with age, your protection against UV radi-ation decreases as you get older. This in turn leads to the appearance of age spots (liver spots). Likewise, grey hairs are also caused by a decrease in the number of melanocytes within the hair follicles.

The dermis is situated below the epidermis. This is where living cells are processed on a continual basis. Unlike the epidermis, the dermis is rich in capillaries that bring fresh nutrients and oxygen necessary for the continued growth of new skin cells. With age, the number of these capillaries decreases, giving rise to the appearance of pale skin, coupled with decreased nutrient flow and less efficient waste removal. This is hampered still further by a diet high in acid-forming foods.

Much of skin ageing involves structural changes in the dermis, particularly to the structural proteins (collagen and elastin) that maintain its flexibility. Wrinkled skin is a combination of reduced collagen synthesis coupled with free-radical damage.

How to Save Your Skin

- Drink more water.
  If water is needed elsewhere in your body, the reservoir that keeps your skin plump and soft will be robbed, so keep it topped up.

- Avoid processed foods.
  Eat an abundance of fruit and vegetables – their high content of antioxidants will zap ageing free radicals.
- Avoid the sun.
  Getting a suntan may help you look great in the short term, but it can do irreparable damage to your skin. Clinical trials have shown that the mineral selenium reduces redness and cell damage, but this is no substitute for avoiding UV rays. These are strongest between 11am and 3pm so cover up and always wear a sunscreen.
- Get a good night's sleep.
  Good-quality sleep is essential because that's when your skin cells go through the process of regeneration and repair. Try to get at least eight hours a night. Don't eat too near bedtime or drink too much alcohol because your body needs to shut down digestion to concentrate on the important business of cell growth and repair. What's more, the enzymes in your liver want to get to work on existing harmful toxins, not to detoxify alcohol.
- Take adequate exercise.
  Too little exercise can result in lacklustre skin. Aerobic exercise is best for your skin. It stimulates your circulation to deliver nutrients to the skin, and your lymphatic system to remove toxins. This aids collagen production, which in turn improves skin texture and moisture retention.
- Cleanse, tone and moisturize.
  Develop a simple bedtime cleansing routine to remove the debris that has collected on your skin during the day. Do not drag the skin, and do not think 'more is better', as you do not want to clog your pores with heavy cream.

For healthy skin you need abundant amounts of vitamins and minerals, particularly the antioxidant vitamins A, C and E and selenium. You also need zinc – stretch marks are a sign of zinc deficiency. The essential fatty acids – omega-6 and, to a lesser degree, omega-3 – are also essential. The first sign of a deficiency of essential fatty acids is dry, flaking skin.

How to Have Great Hair

The secret of healthy hair lies in the food you eat. We feed our animals good food, which in turn makes their coats glossy, and it is the same with us. With a balanced diet, hair will be in tiptop condition. While shampoos and conditioners may help superficially, you need to feed your follicles for good-looking hair all the time.

Hair Structure
The hair is composed mainly of the protein keratin and is therefore a dead structure. The length of the hair is divided into three parts:

- Hair shaft: the part of the hair that lies above the surface of the skin.
- Hair root: the part that is found below the surface of the skin.
- Hair bulb: the enlarged part at the base of the hair root.

Internally, the hair has three layers:

- Cuticle: the outer layer, made up of transparent protective scales that overlap one another. The cuticle protects the cortex and gives the hair its elasticity.
- Cortex: the middle layer, made up of tightly packed cells containing the pigment melanin, which gives the hair its colour. The cortex helps to strengthen the hair.
- Medulla: the inner layer, made up of loosely connected cells and tiny air spaces. The reflection of light through the air spaces determines the sheen and colour of hair.

Hair growth originates in the matrix. Living cells produced in the matrix are pushed upwards away from their source of nutrition. They die and are converted to keratin to produce a hair. The growth pattern of hair ranges from approximately four to five months for an eyelash hair to approximately four to seven years for a scalp hair. Diet, illness and hormonal influences affect hair growth.

How to Feed Your Follicles

- Maintain iron stores.
  Low iron stores are the primary cause of thinning hair in women. Lean red meat, dark-green vegetables and organic eggs are good sources of iron. Vitamin C helps iron absorption.
- Get enough protein.
  We only need a little protein in the diet but, as with all foods on the pH Diet, quality is more important than quantity. Eating a little protein on a regular basis is important for strong healthy hair.
- Get enough vitamin B_3 (niacin) and vitamin B_5 (pantothenic acid). The pH Diet contains plenty of these vitamins so there should be no need to take supplements.
- Eat a varied diet.
  Boring, repetitive diets can lead to thinning hair. By eating a varied diet, you ensure you get the full range of vitamins and minerals required for healthy hair. Aim to eat all 80 alkaline-forming foods within a month!
- Get active.
  Hair follicles receive nutrition from your blood, which circulates more efficiently with exercise.
- Avoid 'dieting'.
  Yo-yo dieting over the years can lead to thinning hair.

Nails

You want your nails to look good because they are on show every day. Unhealthy nails are usually a sign of ill-health. Nails are made of keratin, the same material as the surface of the skin.

Nails grow faster in summer when exposure to ultraviolet radiation increases cell division. A good blood supply is essential to nail growth as oxygen and nutrients are fed to the living cells of the nail matrix and nail bed. Protein, calcium and zinc are good sources of nourishment for the nails. If you are eating a well-balanced and varied diet, it will show in the condition of your nails.

Nail growth may be affected by:
- Ill-health.
  During illness, your nails will receive a reduced blood supply as the body attempts to restore other areas to good health.
- Age.
  Nail growth slows down with age as the blood vessels supplying the nail become less efficient.
- Poor manicure technique.
  Applying heavy pressure when using manicure implements can damage the nail matrix and cause ridges. Depending on the extent of the damage, the ridges may eventually grow out.
- Accidents.
  Shutting a finger in a door, for example, may result in bruising and bleeding of the nail, or even its complete removal. If the nail bed is damaged, the nail could be permanently malformed.
- Diet.
  If a nutrient is lacking in the diet it can result in a diminished blood supply to the nail. The conditions of the fingernails and toenails can be a useful diagnostic aid. The following are some of the more common associations between nail health and nutritional status:

- Absent half-moons   Protein deficiency
- Nail ridges   Protein deficiency, vitamin A deficiency
- Pale nail beds   Anaemia
- Peeling nails   Vitamin A deficiency
- Poor nail growth   Zinc deficiency
- Splitting nails   Sulphur amino acid deficiency (rich in onions and garlic)
- Spoon-shaped nails   Iron deficiency
- Thin, brittle nails   Iron deficiency, calcium deficiency, vitamin D deficiency or hydrochloric acid deficiency (stomach acid)
- Washboard ridges   Iron, calcium and/or zinc deficiency
- White nails   Copper excess or liver disease
- White spots   Zinc deficiency, thyroid deficiency, hydrochloric acid deficiency (stomach acid)

Eyes

Vitamin A plays two indispensable roles in the eye: it helps maintain a crystal-clear outer window, the cornea, and it participates in the conversion of light energy into nerve impulses at the retina. Over 100 million cells reside in the retina, each with about 30 million molecules of vitamin A-containing visual pigments. With good nutrition, vitamin A deficiency can be avoided, and with it age-related problems that affect the eyes.

Eyesight may be affected by:
- Vitamin A deficiency.
- Zinc deficiency – the two enzymes important for retinal function are dependent on zinc.
- Free-radical damage caused by ultraviolet light damaging the fatty acids within the retina.

Protect your eyes by:

- Eating foods high in betacarotene, which will be converted to vitamin A and stored in the liver.
- Eating foods high in zinc – this has an important role to play in the health of the eye.
- Wearing good sunglasses and avoiding direct contact with ultraviolet light.
- Taking a supplement of ginkgo biloba – its antioxidant flavonoids protect the retina of the eye from free-radical damage, preventing macular degeneration.

If you wear glasses, update them regularly. There is nothing more ageing than old-fashioned glasses. While prescription glasses should not be viewed as fashion accessories, trendy glasses can take years off you and be worth every penny.

Vital Nutrients for Healthy Skin, Hair, Nails and Eyes

All vitamins and minerals are important for overall health, but the antioxidants (vitamins A, C and E and selenium) and zinc are especially important for the health of the skin, hair, nails and eyes. A healthy balanced diet, containing as much variety as possible, will provide you with all the essential nutrients you need.

Vitamin A
This vitamin is good not only for the skin inside and out, but also for the urinary tract, the respiratory and digestive systems, and the mucous membranes throughout the body. It is also essential for night vision.

The therapeutic adult dose for vitamin A is 2,000mcgRE (6,000iu). However, if you are pregnant or trying to conceive, do not exceed 3,000mcgRE (10,000iu) or 3,000–30,000mcg betacarotene per day.

RDA (Recommended Daily Allowance): 600mcgRE
Primary sources (per 100g):
Beef liver (35,778iu), carrots (28,125iu), veal liver (26,562iu), sweet potatoes (17,055iu), squash (7,000iu), watercress (4,700iu), cantaloupe melon (3,250iu), cabbage (3,000iu).

Vitamin C
Vitamin C builds up collagen in the skin, prevents scurvy and is reputed to help prevent colds and increase overall disease resistance. Most animals make their own supply but human beings need to ingest it every day.

RDA (Recommended Daily Allowance): 40mg
Primary sources (per 100g):
Broccoli (110mg), peppers (100mg), kiwi fruit (85mg), lemons (80mg), cabbage (60mg), cauliflower (60mg), strawberries (60mg), tomatoes (60mg), watercress (60mg), limes (29mg).

Vitamin E
This vitamin protects cells from damage and helps the body use oxygen. It is reputed to help the production of antibodies and to heal burns.

RDA (Recommended Daily Allowance): 3–4 mg
Primary sources (per 100g)
Unrefined corn oils (83mg), sunflower seeds (52.6mg), wheatgerm (27.7mg), lima beans (7.7mg), tuna (6.3mg), sweet potatoes (4.0mg), sardines (2.0mg), salmon (1.8mg).

Selenium
Selenium assists the actions of vitamin E. Its antioxidant properties help protect against free radicals and cancer-forming substances and reduce inflammation. It also stimulates the immune system and is needed for a healthy metabolism.

RDA (Recommended Daily Allowance): 70mg
Primary sources (per 100g)
Oysters (0.65mg), herrings (0.61mg), molasses (0.13mg), tuna (0.116mg), mushrooms (0.113mg), beef liver (0.049mg), cottage cheese (0.023mg), cabbage (0.003mg).

Zinc
A component of over 200 enzymes in the body, zinc is essential for overall health. This mineral helps bone and teeth formation, keeps hair and nails in good condition, is essential for growth and important for healing. It also promotes a healthy nervous system.

RDA (Recommended Daily Allowance): 15mg
Primary sources (per 100g)
Oysters (148.7mg), ginger root (6.8mg), lamb (5.3mg), egg yolk (3.5mg), oats (3.2mg), rye (3.2mg), whole wheat grain (3.2mg), almonds (3.1mg), haddock (1.7mg), turnips (1.2mg).

Feed Your Face

An ageing skin also indicates an ageing body, as do unhealthy-looking nails and hair. As you and others notice improvements in these areas on the outside, you can be sure that your health is improving on the inside as well. Familiarize yourself not only with all 80 alkaline-forming foods recommended in *The pH Diet*, but also with foods high in antioxidants, and eat them in abundance!

Managing Your Weight

When most people start a new eating programme or diet, they tend to do some of the following things:

- They become overwhelmed by the amount of reading, meal planning and shopping required, and usually sabotage their own best plans.
- They begin doubting themselves: 'Can I really do this?' Doubts and negative thinking do not result in weight loss.
- They start making excuses for their weight: 'It runs in the family'; 'It's in my genes'; 'It's my hormones'. Making excuses does not lead to weight loss.
- They want instant results and don't give the new processes time to work. Feeling impatient, they judge and criticize every step, and bring lots of negative emotions to the process itself. Again, they usually sabotage their own best plans.

All of these behaviour patterns describe how *not* to begin. Doing any of the above will scupper your chances. Some of them will not even allow you to get started. If you continue to play these mind games – as that is what they are – you will achieve nothing positive. You will feel bad about yourself and may need to bury your sorrows in some junk food, caffeine or alcohol.

This book does not go into the psychology of dieting. We can, however, give you a few tips on how to make a positive start to the pH Diet, and indeed any new phase in your life you may be embarking upon:

- Learn to say 'no'.
- Don't make excuses.
- Allow yourself to look great, feel good and enjoy the energy and vitality that can be yours with the pH Diet.
- Be kind and gentle with yourself regarding your current weight.
- Set realistic goals.
- Don't allow yourself to get overwhelmed – you only need to do one small thing every day. Every little thing you do counts.
- Be positive in your thinking – you can *if you think you can*.

Every action starts with a thought. Make your thoughts positive, especially about your food and drink choices.

How Does the pH Diet Help Weight Loss?

The body needs to eliminate the toxins you take in through acid-forming foods and drinks. To do this, it creates fat cells, which carry acids away from your vital organs and try to protect them. In one way fat is saving your life! That is why your body does not want to let it go. When you eat to make your body more alkaline, you will no longer need to keep that fat around. So thinking positive thoughts about nutrient-rich, alkaline-forming foods and *acting upon those thoughts* by eating those foods, will bring success.

The pH Diet assists with weight loss by recommending a healthy amount of alkaline-forming vegetables and fruits. This is particularly the case at Level 3, when you eat fewer processed foods and more raw foods. Eating more vegetables is as central to a healthy nutrition plan as having less sugar and drinking more water.

While almost all vegetables are good, some are far better than others. The general rule is *the greener the better*. Raw fruits and vegetables contain enzymes, and enzymes keep weight down. Like builders who use just enough bricks and mortar to put a house together, enzymes convert the food we eat into the exact quantity of raw materials needed for maintaining and rebuilding the body. The rest is eliminated. However, in the case of enzyme-free food – processed, ready-made or over-cooked food – there is nothing to prevent too many nutrients being absorbed into the body. What the body doesn't need is either stored in fat cells or turned into toxic, acid waste.

Cooked food also causes weight gain because, taking longer to digest, it leaves food particles behind. Leftover food turns into acid waste, some of which is stored in the body as fat. However, by increasing your intake of alkaline-forming food and maintaining an alkaline balance, acid wastes will be greatly reduced and the body will also release retained water. Water retention is a symptom of toxic build-up as the body retains water to dilute the toxins. With fewer toxins to dilute, the body will release the water.

The pH Diet is not a 'diet' in the typical sense. The word diet usually implies short-term deprivation before reverting to old eating habits. This easy, lifelong programme is not just about losing weight. It is designed to help you use food to your advantage – for its nutritional qualities and to give the body what it needs to repair and maintain itself. The pH Diet offers balanced meals across all main food groups. By choosing foods from all main food groups at each meal or snack, you will automatically improve insulin control (*see 'Fighting Fatigue', page 195*). When insulin levels are controlled, then weight loss and weight management are the results.

The pH Diet is not:
- a high-protein diet, but provides adequate protein for the body's building and repairing work.
- a low-fat diet, as it provides essential fats needed for protection and insulation.
- a high-carbohydrate diet, as it is balanced with proteins and fats.

A Balanced Diet

Choosing foods from across the major food groups *at every meal* constitutes a balanced diet. By so doing, you automatically balance alkaline- and acid-forming foods. Protein should account for 15 per cent of daily calories. Most protein foods from animal sources are also acid forming, so by keeping to 15 per cent – one small portion – ideally three times a day, you will be limiting your intake of acid-forming foods.

Remember that fats and grains are also acid forming, but far less than protein foods. Fruits can also be moderately acid forming. Some foods may be alkaline forming when raw but become acid forming when cooked. An ideal balance of alkaline-forming and acid-forming foods would be 75 per cent alkaline forming to 25 per cent acid forming – which is the aim of Level 3. Levels 1 and 2 build up to this.

A BALANCED DIET
Using calories as a measure

ALKALINE- AND ACID-FORMING FOODS

Carbohydrates
55–60% of total daily calories should come from this food source

Take care as some carbohydrates are acid forming, especially white bread, pasta and peeled potatoes. Choose 75% of carbohydrates from the list of 80 Alkaline-forming Foods (quinoa, potato skins, buckwheat) and the other 25% from the 20 Best Acid-forming Foods (brown rice, oats, Burgen bread, Bharti's Bread, rye and corn breads).

Proteins
15% of total daily calories should come from this food source

Choose vegetable sources of protein, such as peas, beans and legumes. As a general rule, you should eat a grain and a pulse together. For example, beans (pulse) on toast (grain) will give you all 8 essential amino acids. Combining rice with lentils increases the protein value by a third.

Fats
25–30% of total daily calories should come from this food source

Fats are also acid forming, so use sparingly. Saturated fats are not necessary for good health and can be eliminated or used in small

- 5–10% saturated fats
- 10% monounsaturated fats
- 10% polyunsaturated fats

amounts. You can cook with olive oil to use your 10% allowance of the semi-essential monounsaturated fats. Use the '3&6 Mix', Essential Balance or Udo's Choice daily for your essential fatty acids (omega-3 and omega-6).

Although percentages can give us a useful guideline, they can be misleading. The most important thing is to keep track of the total amount of fat grams you are eating. We recommend a daily fat intake of no more than 40 grams. To lose weight, the average woman's daily calorie intake should be 1,500 while an average man's should be 1,900. This is dependent upon your activity and energy levels, diet history and level on the plan. While we can guide you, only you know how many calories you really need (not want!).

Do not worry or get overwhelmed by these percentages. It is easier to use the following exercise.

How Do I Know if I am Eating a Balanced Diet

It is not necessary to count calories. Life is too short for such a tedious task. The above table is for information purposes only, so please do not lose sleep trying to work out percentages. To ascertain if your meal is balanced, simply ask yourself these questions every time you eat something. Look at your complete meal, including any drink you may be consuming with it, and ask:

Where is the carbohydrate?
Choose foods from the list of 80 Alkaline-forming Foods, plus a small portion from the 20 Best Acid-forming Foods, such as Burgen bread, Bharti's Bread, rye and corn breads, or kidney beans, lentils, lima beans, oats, brown rice or rye.

Does it make up three-quarters of my plate?
Carbohydrates should comprise 75 per cent of the food on your plate and give you approximately 55–60 per cent of your calories.

Where is the protein?
Choose chicken, cottage cheese, organic eggs, fish, organic meat, turkey, goat's cheese.

Does it take up a small space on my plate?
Only 15 per cent of your meal should consist of protein from acid-forming foods, giving you approximately 15 per cent of your calories.

Where are the fats?
No saturated fat (hard fat like butter and margarine).
Excellent – you don't need it.

Food cooked in a small amount of olive oil, or a little olive oil as a dressing.
Excellent – this provides your intake of monounsaturated fat.

'3&6 Mix' added to a meal or 1 tablespoon of Essential Balance or Udo's Choice.
Excellent – this provides your intake of essential fatty acids.

Your total fat intake makes up the remaining 10 per cent of acid-forming foods.

Explaining the Food Groups

Ideally, you should eat foods from all the major foods groups at every meal. This not only keeps blood sugar levels (insulin) on an even keel, but also ensures the smooth running of all other complex bodily functions.

Carbohydrate

Carbohydrates are your body's number one fuel source. All carbohydrates must be broken down into glucose (blood sugar) to supply your body with energy. Simple sugars (fruit, honey) and double sugars (table sugar, milk) are digested quickly and give your body an energy surge. Complex carbohydrates (starchy and fibrous foods) are digested more slowly. The glucose produced by these different carbohydrate types is used either immediately for energy or stored in muscle cells and the liver as potential energy (glycogen). If there is no room left in the liver and muscle cells, glycogen is converted to fat and housed in your fat cells.

Carbohydrates are the main source of energy for your muscles when you exercise, and since exercise is crucial to healthy living, your body needs carbohydrates. Glucose is also the main source of fuel for the brain.

Protein

Protein exists in almost everything you eat. You will find it not just in meat, poultry, fish, dairy products, beans and nuts – foods that are predominantly protein – but also in some fruits and vegetables.

Why do you need protein? Protein has two main functions, one structural and one metabolic. The protein we eat is broken down, digested and absorbed as amino acids before being rebuilt into all the tissues in your body – hair, skin, internal organs and muscles. That does not mean protein is a magic muscle builder. You strengthen your muscles through exercise, not by eating excess protein. In addition to its structural role in building tissue, protein supports the immune system, manufactures the enzymes, hormones and neurotransmitters that regulate bodily processes and, in the absence of sufficient carbohydrates and fat, serves as fuel for the body. (Carbohydrates, which are more easily digested, are the body's first fuel choice, followed by fat.)

If all protein foods were pure protein, weighing in at four calories per gram, eating large quantities of them would not necessarily be bad. The problem is that many protein foods come bundled with fat, at nine calories per gram. Worse still, the fat tends to be saturated fat. Because of the fat connection, calories in predominantly protein foods are more densely packed than calories in predominantly carbohydrate foods. If your diet is rich in protein foods, you will tend to eat more than you need for your basic fuel requirements. Too much protein will just add to your waistline and, as protein foods are acid forming, will draw upon your body's mineral reserves. Your total daily protein requirement is just 15 per cent of your total calories for that day, so a little goes a long way.

Fat

Fat is an essential component of healthy eating and performs a variety of vital functions. It enables your body to absorb the fat-soluble vitamins A, D, E and K. It plays a role in hormone production and regulation, builds cell membranes, helps keep your skin lubricated, protects your vital organs, keeps you warm and plays a role in the healthy functioning of your immune system. It also gives food flavour and texture. However, not all fats are the same.

Saturated fats tend to be solid at room temperature and are most commonly found in animal products. This type of fat abounds in butter, cheese, battery eggs, red meat and coconut and palm oils. It is notorious for raising harmful cholesterol and promoting heart disease. These fats are not essential and should only make up a tiny proportion of your overall fat intake.

The monounsaturated fats are found in olive oil, olives, avocados and canola and peanut oil. More beneficial to the diet than saturated fats, monounsaturated fats slightly lower harmful LDL cholesterol, while slightly increasing beneficial HDL cholesterol. This is partly because these fats are rich in plant-based antioxidants, which 'mop up' the free radicals that can wreak havoc in the body.

Polyunsaturated fats are prominent in 'seed oils' (pumpkin, sunflower, sesame, linseed), soya, safflower and other cold-pressed vegetable oils. These oils are usually liquid at room temperature. Within this group can be found the omega-3 and omega-6 essential fatty acids. As their name suggests, they are essential for life, and as the body is unable to make them, must be taken in through the diet every day. The '3&6 Mix' (see page 22 and 112) contains just the right amounts of these two essential polyunsaturated fats. You need fat to lose stored fat on your body, and it is the polyunsaturated fats that are associated with this function. This is why so many 'fat-free' diets fail. You need fat in your diet but it is the type of fat you choose that can be crucial.

Polyunsaturated fats should not be heated (the only oil that can be heated safely is olive oil). Sunflower oil and all the other polyunsaturated oils can be used for salad dressings.

What to Look for in a Diet

As this chapter demonstrates, the pH Diet has many benefits including helping you to manage your weight. Other weight-loss diets take approaches that are far less healthy. Before looking at why the pH Diet is such a healthy option for weight loss, we will look at some of these other diets.

Why Counting Calories Does Not Work
Most people live in the fast lane, juggling work, family and a social life. When it comes to wanting to lose a few pounds, people want to do that quickly too. However, the body is very complex. The quicker weight is shed, the quicker it will go back on as a survival mechanism. Slow down! Look at the long-term picture and do not undermine your health by choosing a quick-fix diet.

Unlike the pH Diet, many weight-loss diets are based on calorie counting. We are all familiar with the logic behind this approach: if you eat more calories than you burn off in activity, you gain weight;

and if you eat less than you burn off, you lose weight. Unfortunately, while this sounds good on paper, it does not always work in practice. People on low-calorie diets feel restricted, get hungry and then eat – usually the wrong type of foods.

Another reason why counting calories does not work is that many programmes fail to educate their participants in nutrition. When embarking on a weight-loss diet, all dieters seek ways to cheat the system – this is human nature! So if they are allocated 1,500 calories a day, some dieters may choose less-than-healthy options such as chocolates, expensive 'diet' food and alcohol. The only way to lose weight successfully is to understand food, its nutritional qualities and what it is actually doing for your health.

High-protein Diets

The objective of any weight-loss diet should be to improve health. It is well known that carrying too much weight is detrimental to health, particularly for the heart and arteries. If improving health is the driving force behind any weight-loss diet, why do so many people choose diets that are fundamentally bad for their health?

The popular high-protein/high-fat/low-carbohydrate diets take weight off relatively easily, which is why they are so popular. But at what cost? Being 'allowed' to eat lots of unhealthy foods high in saturated fat – such as bacon, fried eggs, sausages, meat and cream – seems to lessen the hunger pangs that drive many people to binge eating, at least at the beginning of the diet. If the objective in losing weight is to improve health, this is not the diet to choose because it almost eliminates one of the major food groups – carbohydrates. Eating so little carbohydrate deprives the body of glucose, its primary source of fuel, while eating too much protein threatens the body's mineral reserves.

Excessive phosphate levels in meat can remove calcium and magnesium in the teeth and bones. The body needs these minerals to neutralize and buffer the high levels of urea produced when meat is broken down. It excretes them together with the urea.

Any diet that excludes whole food groups should not even be considered if you have any regard for your health. The food groups are there for a reason – they are all needed for specific and important bodily functions. We need a *balance* of the major food groups – the carbohydrates, proteins and fats – together with a *balance* of alkaline- and acid-forming foods.

Comparing Popular Diets

	Carbohy-drates	Essential Fats	Protein	Balanced?	Main Theme
The pH Diet	✓	✓	✓	✓	Balanced across all main food groups. Restores beneficial alkaline reserves to the body.
The Atkins Diet	✗	✗	✓	✗	High in protein and saturated fat, which puts the body into an unnatural state called ketosis, which reduces weight nevertheless.
The carbohydrate Addict's Lifespan Programme	✗	✗	✓	✗	Breaking carbohydrate addictions. Once a day, however, you can eat any amount of carb-rich junk food for one hour!
Weight Watchers	✓	✓	✓	Only if followed *exactly*. Could very easily be unbalanced.	You count the 'points' of everything you eat. There is also a large financial commitment. However, the group meetings are helpful to some.
The Zone Diet	✓ (limited)	✗ The Zone only recommends mono-unsaturated fats.	✓	✗	Basically a high-protein diet, but includes some fruits and vegetables. You have to be rigorous about following a plan and making many calculations.

Balance, Variety and Moderation

Any programme you choose should follow the principles of balance, variety and moderation. No drastic reduction in any food group is healthy over the long term. Be suspicious of overzealous claims that fly in the face of accepted wisdom. If the diet departs radically from the recommendations of the well-documented food pyramid then beware. Current deprivation diets share a disregard for the principles of sensible, lifetime weight control: balance, variety and moderation. Without balance, you could deprive yourself of essential nutrients. Without variety, you could nurture a sense of deprivation that drives you to desperate bingeing and yo-yo dieting. Without moderation, the diet will not work in the long term.

The pH Diet presents an easy, varied, balanced and sensible program. There are many paths to a destination – doesn't it make sense to opt for one of the healthiest?

Planning Principles

As well as balance, variety and moderation, the pH Diet is also based on the following diet-planning principles. Before you begin the programme, you should read through them and take them into consideration.

B A L A N C E

Balance
Adequacy
Liquid
Abundance
Nutrient density
Calorie control
Empty calories

Balance
The art of a balanced diet involves consuming enough of each food type. For the average person, 55–60 per cent of their diet may comprise carbohydrate foods, 15 per cent protein foods and 25–30 per cent fats. For optimum health, your intake of vitamins and minerals should be balanced as well. For example, it is important to get a balanced amount of the essential minerals calcium and iron. Meat, fish and poultry are rich in iron but poor in calcium. Conversely, milk and milk products are rich in calcium but poor in iron. For a balance, choose some meat or meat alternatives for iron and some milk or milk alternatives (soya) for calcium.

Try to balance your intake of fats so that you are consuming monounsaturated, polyunsaturated and perhaps a little saturated fat. The '3&6 Mix' (*page 112*) will enable you to balance your intake of omega-3 and omega-6 fatty acids without any difficulty. Another important balance is, of course, between alkaline- and acid-forming foods.

Adequacy: Getting What You Need
Adequacy means that the diet provides sufficient energy and enough nutrients to meet the needs of healthy people. Each day the body loses some iron, for example, which needs to be replaced by iron-containing foods. A person whose diet fails to provide enough iron may develop symptoms of iron-deficiency (anaemia). Each day, adequate nutrients must be supplied by the diet, particularly the essential amino acids, essential fatty acids and water. Choosing a variety of foods will ensure adequacy of vitamins and minerals.

Liquids
Water is necessary and vital for good health. When considering a new dietary programme, it is important to establish the recommended fluid intake. Any diet recommending unlimited black coffee, tea, 'diet' or fizzy drinks does not have your health in mind.

Abundance

Is there an abundance of fresh fruits and vegetables in your diet? You must eat at least five servings of fruits and vegetables every day. Fruits and vegetables supply the vitamins and minerals that not only keep us healthy but also provide the second part of many enzymes – the co-enzymes – which are responsible for hundreds of chemical actions in the body every minute of every day.

Nutrient Density

Nutrient-dense foods deliver the most nutrients and fewest anti-nutrients for the least amount of calories. Consider foods containing calcium, for example. You can get about 300 milligrams of calcium from either 100 grams of Cheddar cheese or 1 cup of non-fat milk, but the cheese contributes about twice as much food energy (calories) as the milk. The non-fat milk, then, is twice as calcium-dense as the cheese; it offers twice the amount of calcium for half the calories.

Calorie Control

The average woman requires 2,000 calories, and the average man needs 2,400 calories, per day to maintain their weight. To lose weight, a woman needs 1,500 calories and a man requires 1,900 calories per day. Designing an adequate, balanced diet without overeating requires careful planning. The key to calorie control is choosing food with a high nutrient density.

Empty Calories

Beware of diets that recommend 'empty' calories. A glass of cola and a bunch of grapes each provide about 150 calories but the grapes offer a trace of protein, some vitamins, minerals and fibre along with the energy. The cola, on the other hand, offers only empty calories from sugar without any other nutrients. A diet recommending unlimited low-calorie drinks, 'diet' meals, reduced-fat or no-fat products, or processed foods does not have your health in mind.

Does the pH Diet meet the requirements of the BALANCE test? Yes, the pH Diet offers you all the above and more.

Fighting Fatigue

Britain's biggest ever health survey, conducted in 2001 and involving 22,000 people, found:

- 76 per cent of people are often tired
- 58 per cent suffer from mood swings
- 52 per cent feel apathetic and unmotivated
- 50 per cent suffer from anxiety
- 47 per cent have difficulty sleeping
- 43 per cent have poor memories or difficulty concentrating
- 42 per cent suffer from depression

The majority of our clients are always tired. They wake up tired, are tired all day and go to bed tired. They cannot remember when they last felt energetic or full of life. Some of these clients are teenagers and others are in their 20s and 30s when high energy levels should be normal, not a distant memory. How you think and feel is directly affected by what you eat, and how much energy you have is also affected by your choice of foods.

The Problem with Sugar

The biggest factor in fighting fatigue is sugar. Sugar dulls the mind and has a huge impact on insulin production. Over the years, this can result in insulin resistance, accompanied by constant fatigue.

We have been advised over and over again that carbohydrates, especially complex carbohydrates, are good for us; that they will give us energy because they are the main source of fuel for the body and brain. This is all true. However, somewhere along the line we started eating more and more carbohydrates, hoping for ever-increasing levels of energy. At the same time, we began to consume less fat in the hope of losing weight. This combination leads to an unbalanced diet.

When you eat any carbohydrate, your body releases insulin. When you eat lots of carbohydrate, your body releases lots of insulin. The cells eventually become resistant to the insulin and cannot receive the glucose they need to give you energy. The result is fatigue. It must be remembered, therefore, that all carbohydrates – whether complex or simple, whether 'good' or 'bad' – are all broken down to the same substance before absorption can take place. That substance is glucose.

What is Insulin?

Insulin is a hormone produced by the pancreas. You need it to live but you probably have far too much insulin floating around in your body. Most adults have about one gallon of blood in their bodies and are quite surprised to learn that in that gallon, there is only one teaspoon of sugar. You only need one teaspoon of sugar at all times, if that. If your blood-sugar level were to rise to one *tablespoon* of sugar, you would quickly go into a hyperglycaemic coma and die.

Your body works very hard to prevent this by producing insulin to keep your blood sugar at the appropriate level. This reaction keeps you from dying when you eat sugar. Unfortunately, it turns out that high levels of insulin are quite toxic for your body. Any time you eat any kind of sugar – be it fruit, honey, refined foods (white bread and rice), sweetened breakfast cereals, alcohol, biscuits and cakes and so on – you are increasing your insulin levels. If you have high cholesterol levels, high blood pressure, type II diabetes or are overweight, it is highly likely that you are eating far too many carbohydrates.

While too much insulin accelerates ageing and fatigue, bear in mind that without enough insulin your cells will starve to death. You can push insulin too low by eating a high-protein, low-carbohydrate diet. This is why it is so important to have a balanced diet that includes all the major food groups *with each meal*.

Why Do I Put on Weight When I Eat so Little Fat?

One of insulin's jobs is to enable your body to make use of glucose by escorting it into the cells. Here it is burnt to give you energy. If there is too much glucose in your blood, however, insulin turns some of it into glycogen, which is stored in your muscles and liver. This glycogen can be turned back to glucose for energy when you need it. If there is still glucose in the blood after the stores of glycogen in the muscles and liver are full, insulin converts it to fat and stores it where it cannot do any harm – in the fat cells. This is what happens if your diet is consistently too high in carbohydrates with insufficient protein and fat.

Improving Glucose (Sugar) Tolerance

Glucose is the simplest chemical form of sugar. As we have seen, all the carbohydrate (sugar and starch) you eat is broken down by the digestive system into glucose – the only form in which it can be absorbed by the body and turned into energy.

The glucose enters the bloodstream as soon as digestion is complete. Normally, the pancreas then reacts by producing insulin, which takes the glucose out of the blood and into the cells.

What Goes Wrong?
If you eat sugar regularly, the pancreas is constantly stimulated. If you eat *any* carbohydrate in refined form (white sugar, sweets,

chocolate, white flour), digestion is rapid and glucose enters the blood in a violent rush. In each case, the pancreas can over-react and produce too much insulin. Blood glucose then takes a rapid, uncomfortable drop and may end up too low for normal functioning (hypoglycaemia). If this over-stimulation happens too often, the pancreas becomes exhausted. Now, instead of too much insulin, it produces too little. Too much glucose remains in the blood (hyperglycaemia). In its most severe form, this condition becomes diabetes.

The regulation of blood sugar is a constant balancing act. The aim is to provide energy to the cells that need it (including the brain) and to make sure that unwanted glucose is not left circulating in the blood. If this balance is lost, both physical and mental wellbeing are, in turn, unbalanced. Low blood glucose (hypoglycaemia) and high blood glucose (hyperglycaemia) can have similar and wide-ranging effects:

- irritability and aggressive outbursts
- nervousness
- depression and crying spells
- vertigo and dizziness
- fears and anxiety
- confusion and forgetfulness
- inability to concentrate
- fatigue
- insomnia
- headaches
- palpitations
- muscle cramps
- excessive sweating
- digestive problems
- allergies
- blurred vision
- lack of sex drive

What Can the pH Diet Do?

There are many things you can do to prevent fatigue, and the long list of symptoms listed above. Following the pH Diet is one of them. In doing so you may also avoid developing type II diabetes (adult onset), which is becoming more and more common among younger people. So cutting down on stimulants (Level 1) and refined carbohydrates (Level 2) will greatly help regulate glucose metabolism.

Vitamins (especially B and C) give important support to the adrenal glands while things get back to normal. Chromium is important in the formation of glucose tolerance factor (GTF), a substance released by the liver that makes insulin more effective. Because the pH Diet includes a little protein with each meal, and encourages the intake of daily essential fatty acids, glucose intolerance will most likely correct itself in time. If you follow our programme, you do not have to worry about any of this. As the diet is balanced across all major food groups, it will automatically regulate glucose metabolism.

How to Fight Fatigue
- Eat small, regular meals, preferably containing protein.
- Always eat breakfast.
- Avoid sugar, and foods containing sugar.
- Avoid concentrated fruit juices, or dilute them with water. Check labels carefully as many fruit juices have added sugar.
- Avoid foods containing preservatives.
- Avoid convenience foods – they are almost certain to contain refined carbohydrate and various harmful chemicals.
- Eat less dried fruit.
- Avoid alcohol.
- Avoid tea and coffee. Decaffeinated coffee is also best avoided as it contains other stimulants.
- Avoid (or cut down) on cigarettes.
- Take regular exercise.
- Do all you can to avoid stress.

Why Deep Breathing Is Good For You

We do not think about breathing because it is automatic, but breathing properly is essential for good health. The kidneys and lungs are our main organs of detoxification. Deep breathing is good for you, getting the oxygen into every cell of your body and the waste material – carbon dioxide – out.

The main purpose of breathing is to get oxygen to the cells. Cells need oxygen in order to function. They also produce carbon dioxide, a natural waste product. If the cells are full of that waste, there is not enough room for new oxygen. The carbon dioxide must therefore be eliminated properly.

The purpose of breathing, then, is not only to bring oxygen into your body but also to dispose of the carbon dioxide. Your blood cells can carry the greatest amount of oxygen when your internal environment is slightly alkaline. It is important that the carbon dioxide be disposed of as quickly as possible as it is toxic waste material and must be expelled from the body.

Chronic hyperventilation or 'over breathing' can make the body too alkaline. This condition is associated with nutrient imbalances, such as low levels of the mineral magnesium. Foods can also affect breathing patterns, possibly due to their effect on the acid–alkaline balance of the body chemistry.

Alleviating the Symptoms of Arthritis

Arthritis, in its many different forms, is the most common form of degenerative disease. Diet plays a large role in its prevention and management.

Broadly speaking, there are two main types of arthritis. Both result in joint pain and inflammation yet have completely different mechanisms and causes.

Osteoarthritis

In osteoarthritis, degeneration begins with the joint itself. This is usually triggered by a calcium deficiency within the bones or a previous physical injury. The disease spreads outwards, causing inflammation in the surrounding area.

Maintaining a proper calcium balance is critical for sufferers of this condition. You want the calcium in your diet to be used by your bones and joints, not to neutralize the acid-forming foods you are eating. The best way to get your calcium is from alkaline-forming green leafy vegetables, and the non-dairy sources of calcium (*see page 168*).

Calcium balance is regulated by calcium, magnesium and vitamin D. Exercise also improves calcium balance, while obesity, bad posture and hard physical labour may contribute to the problem by putting further stress on weight-bearing joints.

Rheumatoid Arthritis

In rheumatoid arthritis, joints can fuse together and a distorted build-up of calcium can cause enlarged joints. Rheumatoid arthritis is not a disease of the skeletal structure like osteoarthritis, but of the immune system, known as an autoimmune disease. White blood cells within the immune system are designed to track down any outside invaders and destroy them. In the case of an autoimmune disease, they identify parts of the body as foreign and attack them by mistake. In rheumatoid arthritis, the joints are attacked.

People with rheumatoid arthritis often suffer from allergies. Hormonal and prostaglandin imbalances may make the inflammation worse. Too much copper and iron, and deficiencies of zinc or manganese also exacerbate the problem.

Gout

This type of arthritis affects the fingers and toes, particularly the big toe. It is due to a build-up of uric acid crystals.

Because purine (a white crystalline compound) and nucleic acid break down into uric acid, doctors recommend that individuals with gout avoid eating foods with a high content of purine and/or nucleic acid, such as liver, sweetbreads, game, herring, anchovies, lobster, crab, sardines, pork and avocado. With the possible exception of pork, however, foods rich in purine and nucleic acid are not foods we tend to eat every day. Foods not eaten on a regular basis are not apt to cause gout, so there is no reason why gout-prone individuals cannot eat these foods every so often.

It is particularly unwise to avoid sardines and avocados because the abundance of nucleic acids in these two foods helps regenerate body cells. People living on the coast of Portugal who work in the sardine industry and eat sardines every day are famous for their youthful appearance and robust health in old age.

Individuals with gout invariably develop osteoarthritis. There is, however, one important difference. Gout is caused by an excess of uric acid only, and osteoarthritis by a variety of acid wastes. The difference between these two kinds of arthritis is in the nature of the acid waste deposited in and around bones and in muscles. Whether individuals develop gout or osteoarthritis depends on what they eat and in which enzymes they are deficient.

How the pH Diet Can Help

A supplement containing enzymes that break down fat, protein and carbohydrate can often help by improving digestion. Good digestion eliminates, to some extent, the acid waste by-products of undigested food debris that end up in mineral deposits on bones, tendons and muscles.

Calcium is relatively well absorbed by the body. On average, 30 per cent of ingested calcium reaches the bloodstream. Its absorption, however, depends upon many factors. An excess of alcohol, a lack of stomach acid or an excess of acid-forming foods *decreases* its absorption, as does the presence of lead. Once calcium is in the body, there are many factors that influence calcium balance. Sodium, tea, cocoa and red wine make calcium less retainable.

The pH Diet could almost have been written with arthritis in mind. Its abundant vegetables, '3&6 Mix', adequate protein and organic sulphur-rich eggs play a large part in reducing the inflammation of arthritis and all other types of inflammation in the body.

Resources

Suppliers

Bharti Vyas Beauty Centre
24 Chiltern Street
London W1M 1PF

Tel: 020 7486 7910
www.bharti-vyas.com

The Ludlow Clinic
3 Parkway
Corve Street
Shropshire SY8 2PG

Tel: 01584 878427
www.theludlowclinic.co.uk

Supplies: Hydrion pH testing
strip paper; '3&6 Mix' (a mix of
organic flaxseeds, pumpkin,
sesame and sunflower seeds in a
balanced ratio); Osteoporosis
Risk Assessment Test; personal
or postal nutrition consultations
in Ludlow or London

Enzyme Process
4 Broadgate House
Westlode Street
Spalding
Lincolnshire PE11 2AF

Tel: 0845 1300 776
Fax: 01775 761104
e-mail: info@enzymepro.com

Supplies: Alkalizing
supplements: *EnzymePro* 'Body
Balance' (alkalizing minerals
from a special blend of deep
green organic food extracts);
digestive enzymes: *EnzymePro*
'Digeszyme V' (vegetarian
enzyme complex, containing
amylase, protease, lipase and
cellulase: works in pH range
3.5–8.5); fresh water filters;
Hydrion pH testing strip paper

BioCare Limited
Lakeside
180 Lifford Lane
Kings Norton
Birmingham
B30 3NU

Tel: 0121 433 3727
Fax: 0121 433 3879
e-mail: biocare@biocare.co.uk

Supplies: Alkalizing supplements:
BioCarbonate (sodium and
potassium bicarbonate for
balancing intestinal pH);
digestive enzymes

The Nutri Centre
7 Park Crescent
London W1B 1PF

Tel: 020 7436 5122
Fax: 020 7436 5171
e-mail: sales@nutricentre.co.uk
www.nutricentre.com

Supplies: Freeze-dried wheatgrass; Hydrion pH testing strip
paper; digestive enzymes;
Essential Balance; Udo's Choice
(a liquid that provides a perfect
balance of both omega-3 and
omega-6 essential fatty acids, as
well as other important fatty
acids such as GLA – keep
refrigerated and do not heat)

Practitioners

Suzanne Le Quesne Dip ION
Clinical Nutritionist, Author,
Lecturer and Educator on pH
issues

The Ludlow Clinic
3 Parkway
Corve Street
Ludlow
Shropshire SY8 2PG
Tel: 01584 878427

and

**The Institution for Optimum
Nutrition**
13 Blades Court
Deodar Road
London SW15 2NU

Peter Bartlett DO, ND
Naturopath and Osteopath,
Lecturer on pH and acid-base
issues
126 Harley Street
London W1G 7JS
Tel: 020 7935 2030

and

Latymer
Halls Lane
Waltham St. Lawrence
Reading RG10 0JB
Tel: 0118 934 4203

Useful Websites

www.bharti-vyas.com
Profile your Ayurvedic constitution and learn how you can use the
Bharti Vyas method to boost your inner and outer well-being and
beauty.

www.theludlowclinic.co.uk
The Ludlow Clinic is run by Suzanne and Barry Le Quesne.
On the website you will find weekly recipe ideas and motivational
tips. Your questions will be answered personally or you will be
directed to a page of frequently asked questions.

www.nutricentre.com
For books and nutritional and herbal supplements.

Index

Make
www.thorsonselement.com
your online sanctuary

Practitioners

Suzanne Le Quesne Dip ION
Clinical Nutritionist, Author,
Lecturer and Educator on pH
issues

The Ludlow Clinic
3 Parkway
Corve Street
Ludlow
Shropshire SY8 2PG
Tel: 01584 878427

and

The Institution for Optimum Nutrition
13 Blades Court
Deodar Road
London SW15 2NU

Peter Bartlett DO, ND
Naturopath and Osteopath,
Lecturer on pH and acid-base
issues
126 Harley Street
London W1G 7JS
Tel: 020 7935 2030

and

Latymer
Halls Lane
Waltham St. Lawrence
Reading RG10 0JB
Tel: 0118 934 4203

Useful Websites

www.bharti-vyas.com
Profile your Ayurvedic constitution and learn how you can use the
Bharti Vyas method to boost your inner and outer well-being and
beauty.

www.theludlowclinic.co.uk
The Ludlow Clinic is run by Suzanne and Barry Le Quesne.
On the website you will find weekly recipe ideas and motivational
tips. Your questions will be answered personally or you will be
directed to a page of frequently asked questions.

www.nutricentre.com
For books and nutritional and herbal supplements.